The **Strings** *of* **God**

March 2015
Gary Marlin Sandquist

Our Universe as created from the Big Bang

13.8 billion years ago, our Universe suddenly burst into existence from a cosmological event known as the Big Bang when space and time erupted into the spectacular display we now reset call our Universe.

Cosmologists now believe they can correctly account for the entities we observe and measure throughout our Universe.

The composition, distribution and behavior of these entities, namely the stars planets and galaxies we observe, can be described and modelled by known physics. However, there remain perplexing observations that are beyond our present understanding including Dark Matter and Dark Energy.

Copyright © 2015   Gary Marlin Sandquist
ISBN: 10:1460978994
ISBN-13:978-1460978993

i

Cosmic Microwave Background (CMB) from **European Space Agency Probe**

CMB radiation shown 380,000 years after Big Bang

The initial charged plasma formed hydrogen and helium revealing a present dim microwave photon glow at 2.7 K.

Our Universe is now expanding at 67.3 kilometers per second per megaparsec (Hubble constant)

Estimate of Universe composition

68 % Dark Energy
27 % Dark Matter
5 % Normal Matter

DEDICATION

The Strings of God is dedicated to those who labor with their religious beliefs in God and the astounding new findings and developments in modern science, particularly with astronomical investigations and cosmology.

Science recognizes the credibility of the Anthropic Principle that the universe is finely tuned to permit life and humans to exist. The existence of the genetic code imbedded with the DNA genome of life.

This literary work is the effort of the author to provide his own assessment of the current state of science and religion and how this assessment conforms to his rational and spiritual consciousness.

The

STRINGS *of* GOD

A Science Based Novel of Cosmology & Religion
by
Gary Marlin Sandquist

TABLE OF CONTENTS

ACKNOWLEDGMENTS

The author acknowledges the US Government and in particular to the National Aeronautics and Space Administration (NASA) for their support and funding of the essential scientific studies and physical systems used to explore our Universe.

The Author employed many of the resources provided by US NASA on its excellent web site: http://www.nps.gov/chcu/photos.

The Strings of God

By Gary Marlin Sandquist

David ben-Steinmann is a young theoretical cosmology physicist

David	דָּוִד
ben-Steinmann	ben-Steinmann

CHAPTER 1

AAAS Conference at Rensselaer (1 Oct Friday)

David ben-Steinmann, the Weisenhausern Endowed Professor of Theoretical Physics at Harvard, has been invited to present his recent theoretical developments in "M-Theory" for the American Association for the Advancement Science (AAAS). More commonly known as the AAAS, these assembled scientists will hear David's presentation at the closing Grand Council Meeting to be held at Rensselaer University in Albany, New York. The Grand Council Closing Conference culminates the 2010 Annual Conference of the AAAS and traditionally sets the stage for future research goals for major advancements in Theoretical Physics anticipated during the following year. This closing Conference for 2010 will provide David with the opportunity to present his latest mathematical developments and projected experiments. David believes his proposals will lead to a defensible demonstration that additional, spatial dimensions do exist and can be experimentally confirmed. If his pioneering mathematical developments are successful and supported with experimental confirmation, M-Theory will be accepted as the most promising candidate for the final scientific merger of the smallest and largest constituents of energy and mass in the universe. The merger, if successful, will unify general relativity and quantum mechanics that are presently incompatible. Furthermore, the shroud that separates dark matter and dark energy from our visible universe will be

1

pierced and progress can be made in unifying present physical theory with these dark, unresolved entities. This Conference will be a critical event in David's crusade to achieve acceptance for M-Theory that has been the focus of David's brilliant career in cosmology. David has devoted the last two years in preparing the basis for his presentation during this Grand Council Conference of the AAA. He must insure that his presentation is accurate and compelling. The presentation must garner the support required to grant his request for approval and execution of critical experiments on the Large Hadron Collider near CERN, Switzerland to corroborate his hypotheses.

David prepares for AAAS Conference (1 Oct Friday)

David leaves a message at the desk of the Supervisory Emergency Ward Nurse for his sister Rebecca, a registered nurse at Massachusetts General Hospital, to confirm that she can drive him to Rensselaer for his presentation tomorrow. David emphasizes in his recorded message that they must leave no later than 2:00PM in the afternoon from the apartment they share in Belmont, Massachusetts to make the 160 mile, 3-hour trip to Albany and provide him sufficient time to set up his presentation and acclimate to the conference auditorium where he will speak. The scheduled time for his presentation is 6:00 PM, and it is anticipated that this final AAAS Grand Council Meeting will attract an over flow crowd of national and international scientific attendees.

Of intense focus for David's presentation is the large contingency of astrophysicists and cosmologists who will be attending from the Large Hadron Collider or LHC near CERN. The LHC has recently come on line and the operating staff is preparing the next year's schedule of major experiments to be conducted on this unique, world-class, high-energy physics machine. Experimental run time on this multibillion dollar, scientific wonder is very limited and greatly debated by the LHC Experimental Review Board for the rationed machine time available. Therefore, David must persuade the AAAS attendees at the 2010 Annual Conference, many of whom are members of the New Experiments Review Board, that his proposed experiments should be scheduled for performance. David's speech must also persuade all the scientists at the LHC attending the AAAS Conference to acknowledge and support the extended series of difficult experiments he will propose for confirmation of his hypotheses regarding M-Theory and its ramifications for physics and especially cosmology.

David also realizes that Rebecca must return that same evening from the AAAS Conference in Albany, New York to Massachusetts General Hospital where she is an Emergency Ward nurse and has the midnight shift. This early return concerns David because of the time conflicts, but it is essential to have Rebecca transport him to Albany for the Conference. David's Physics Department Chairman was the alternative driver for transportation to Albany, but he is unavailable now because of prior commitments.

Rebecca will drive David to Conference (1 Oct Friday)

As David closes his message to Rebecca, the Chair of the Physics Department enters David's office at Harvard and inquires if David has everything he needs for the AAAS Grand Council Conference. The two sage faculty members sit together in David's Office that is filled with open books and journals, notes, and the other tools used by intense researchers in physics. They discuss the AAAS Conference and David's strategy for gaining support from the LHC Community for David's experimental proposals. The Chair conveys his regret that he cannot drive David to Albany because of his prior meeting with the Board Members reviewing David's Endowed Chair funded by the Weisenhausern Trustees. David acknowledges the Department Chair's commitment to insure that David's endowed appointment at Harvard remains fully funded. He informs the Chair that his sister Rebecca is available, and she will drive him to Albany.

Upon departure of his Department Chair, David rapidly prepares to leaves his office at Harvard by gathering his Apple Laptop Computer and his

Power Point presentation that he has spent the last three weeks preparing for the AAAS Conference. He downloads a complete file with his presentation onto two 4.7 Gigabyte DVDs and a 32-Gigabyte Flash Drive for backup in the event his Laptop Computer fails to operate properly. As he scrolls through the numerous files loaded on the DVD's, he reconfirms those equations and results he will emphasize for the Conference. He believes that his comprehensive mathematical derivations and recent results indicate that at least one higher dimension, beyond the standard four spacetime dimensions in General Relativity, will exhibit an ephemeral, but measurable lifetime during the intense proton beam interactions and collisions at the 7 TeV, (7 teravolts or 7,000,000,000,000 eV), energy range on the LHC. David's analysis indicates that definitive evidence of this higher dimension will be the measurable loss of total energy of the collision products for certain particle specie as their energy rapidly dissipates into higher dimensions above the four spacetime dimensions. David's hypothesis is that the total, virtual and measurable energy is conserved in spacetime for times comparable to the proton beam interaction event at the 7 TeV energy level. The transient energy dissipation should transfer some measurable energy into one or more dimensions that are external to and outside our four-dimensional spacetime universe. Admittedly, this very short time and large energy measurement is extremely difficult and illusive to detect and will require special measurement procedures and expensive, new equipment to perform and confirm the experimental measurement. David has estimated from his mathematical analysis that the time interval for these particle interactions and collision distributions will only be a few femtoseconds (10^{-15} s). David muses that this extremely short time byte is about the time duration for a light photon travelling at the speed of light or 300 million meters per second to traverse a single water molecule. The travel distance is comparable to the wavelength of visible light...only few nanometers.

With his timeworn, aluminum Samsonite Brief Case loaded with the necessary materials for the AAAS Conference, David quickly crosses the Harvard campus compound, exits the front gate and crosses to the station for the Massachusetts Bay Transport Authority referred to as the MBTA by Boston residents. A crowded passenger trip brings David to his regular exit terminal where he climbs the MBTA Belmont exit escalator and makes a rapid transit to his two-story apartment in Belmont, Massachusetts.

As David enters the apartment, Rebecca is waiting for him and has set out the suit that she emphasizes he must wear for his presentation at the Conference. David dislikes suits and especially neckties and complains about the formality and uncomfortable clothes he is expected to wear at these Scientific Meetings. However, Rebecca reminds him that he must present himself to the Conference attendees as a "mature and compliant physicist" in spite of his youthful age. Rebecca cautions David that most of his professional critics are older and disdain the lack of appropriate attire worn by the younger, iconoclastic company of physicists that David represents. David grudgingly acknowledges Rebecca's assertion and reluctantly complies after declaring that he will hang the necktie around his open collared, button-down dress shirt and complete the cumbersome, tie-tying protocol while he is in the car traveling to the AAAS conference.

David inquires if Rebecca has brought his anti-seizure medicine and the syringes she may need to deliver his prescribed medication if necessary. Rebecca sarcastically retorts that she is a competent, registered nurse at Massachusetts General Hospital and well acquainted with David's medical needs. She reproaches him, responding that she is always prepared for his medical exigencies. Rebecca scowls at David and declares sharply,

"Yes, dear brother, I will have my usual medical supplies for your care. However, please do not exhibit a seizure during our trip. I do not have the time or patience this evening to treat you. Remember, David, I must

5

return to Massachusetts General Hospital for my night shift on time. I was late for my evening session earlier this week."

David grimaces and reluctantly informs Rebecca he will terminate the usual extended discussions following the lecture for her to meet her shift schedule at the Hospital.

David and Rebecca travel to Albany (1 Oct Friday)

David and Rebecca rapidly load David's 2007 Honda Accord; and with Rebecca driving and David in his usual location in the front passenger seat; they begin the trip to New York. Rebecca skillfully weaves the Honda through local, congested Belmont traffic and quickly enters the I-90 West Interchange. Boston traffic traveling west from Boston along I-90 even at 2:15 PM in the afternoon is congested and sluggish as expected. However, Rebecca has extensive experience at negotiating this merging interstate labyrinth, and the residual local traffic soon diminishes as she and David depart the metropolitan area along the I-90 Interstate towards New York.

It is early October, and the moderate rain shower predicted for that day will not arrive until evening. However, weather along the New England Coast can be capricious and surprise even seasoned meteorologist's predictions. As Rebecca drives to Rensselaer, she is silent as usual to allow David to concentrate on his presentation and review final details for his presentation. She realizes the importance of this International Conference for David's research.

David occupies the passenger seat with his 17-inch screen, Apple Laptop Computer resting on his knees while he reviews his Power Point presentation. He concentrates on the several slides that outline the experiments he will propose to the audience, especially to the LHC contingent from CERN. David is hopeful these experiments will provide initial experimental evidence of additional spatial dimensions. David's recent solutions to the complex set of M-Theory tensor field equations indicate that additional dimensions will be evident during the ephemeral, femto-second particle lifetimes for the burst produced in the colliding beams of protons when accelerated to at least 7 TeV.

David and Rebecca arrive at Conference (1 Oct Friday)

Rebecca transports David to Albany, New York without incident and limited verbal exchange. The Jewish siblings complete the 160-mile journey along I-90 and arrive into the suburbs of Albany, New York. David probes the GPS on his I-phone and provides Rebecca with the final driving directions to the Marriott Hotel hosting the AAAS Conference. Rebecca carefully follows David's instructions and soon parks David's 2007 Honda Accord in the valet parking area at the Marriott Hotel. She assists David gather his papers spread throughout the car, and he carefully places them in his briefcase. Rebecca passes the car keys to the valet attendant who has been awaiting David's arrival. David and Rebecca quickly enter the reception hall at the Hotel. They are greeted by one of the Conference Organizers who recognizes David and immediately directs them to the Marriott Hotel's Grand Auditorium where David will make his presentation. The Auditorium is already partially occupied by guests as the hotel staff completes final audiovisual preparations for the pending closing Conference. David places his laptop computer on the rostrum, powers up the computer and opens the Power Point presentation from his computer desktop file. He then projects the title slide for his presentation and carefully focuses the image on the screen. Satisfied with the slide focus, David closes the Power Point file and returns his file to the computer desktop inserted into the file Icon labeled, Final AAAS Presentation, David ben-Steinmann.

While David is preparing and confirming operation of the equipment and details for his presentation, Rebecca quickly goes to the Conference Registration Desk to inform the Conference Leaders that their featured speaker for the Grand Council Conference has arrived and is prepared to present the featured Closing Paper. Rebecca is welcomed by the staff who are well acquainted with her and the important role she performs to support

David at such scientific conferences. Rebecca gathers the Conference Identification Badges for both David and herself and several copies of the Conference Program. She quickly reviews the Program and observes that David's presentation is correctly identified in the Program.

David completes the setup for his presentation and retests the operation of the microphone and the projector for the Power Point Slides. Satisfied with all the details for the presentation, he then joins Rebecca at the Conference Desk in the Lobby of the Hotel. David confers with several Senior AAAS committee members who express curiosity regarding his presentation. David adroitly informs them they must wait to witness presentation with all the other attendees for the closing session, but he promises it will be worth their time to attend his presentation.

Rebecca accompanies David to most of his scientific conferences, which the Massachusetts General Hospital Administration condones because of David's preeminence in the field and his association with the Harvard Faculty. Usually, the Hospital can accommodate to Rebecca's absences, but this evening she must return to cover her evening shift at the Hospital because of the limited evening nursing staff available tonight.

Besides serving as a Registered Nurse at Massachusetts General Hospital, Rebecca also acts as the close companion and technical assistant for David during his important scientific meetings not held at Harvard. David is very dependent upon Rebecca for transportation and assistance with last moment details that are essential for David's successful presentations at such scientific meetings. Although Rebecca realizes that David takes her services for granted, she is grateful for this diversion from her routine duties at the Hospital. She also enjoys the vicarious recognition she receives from notable scientists because she is the sister of David ben-Steinmann, the Endowed Professor of Physics at Harvard.

More important, Rebecca is a Registered Nurse and knowledgeable and experienced in managing David's occasional epileptic seizures that can occur without warning at any time. So she is always seated near David at such meetings in the unpredictable and rare event that a sudden seizure occurs. Most of the attendees at these Scientific Conferences are also aware of David's epileptic tendencies and respect David's medical restrictions. The attendees recognize the need for Rebecca's immediate presence. She can quickly recognize a seizure onset and administer the necessary drugs prescribed to control the seizure and prevent David from harm to himself and others. Fortunately, his carefully prescribed drugs have been effective in controlling his seizures, and Rebecca is able to quickly assess the symptoms associated with the seizure and control the episode with rapid administration of the proper drug.

Although epilepsy poses increased responsibilities for David's physical surveillance in addition to Rebecca's duties at Massachusetts General Hospital, she recognizes and enjoys her peripheral recognition for her support of David. She has met most of the preeminent people in science, particularly those in cosmology and astrophysics. To be the sister of David ben-Steinmann provides Rebecca with recognition and social experiences far beyond those arising from her routine nursing position at Massachusetts General Hospital garners.

Rebecca asks David if everything is prepared for his presentation; and David says,

"I have confirmed the satisfactory operation of the equipment and final preparation for my Lecture. I am disappointed that we cannot attend the Post Conference Dinner with the Conference Attendees. Several of the Group Leaders at LHC will also attend the Dinner, but I realize that you must return to Boston this evening to assume your nursing shift. So I have informed the hosts that we must return to Massachusetts General Hospital immediately after my Grand Council Closing Conference presentation."

Rebecca responds curtly saying,

"Thank you, David."

David's Conference Presentation (1 Oct Friday)

David and Rebecca gather in the auditorium and prepare for the Opening of the Conference. After the usual formalities initiating the AAAS Grand Council Closing Conference, the moderator for the final Conference Session is introduced. The moderator for this AAAS Grand Council Closing Conference is the President of the National Academy of Science or NAS. He welcomes the audience to the closing conference and then introduces David with adulatory praise. The President gives a brief account of David's extraordinary accomplishments in M-Theory and his eminence in cosmology. The President asserts that he is familiar with David's recent solutions for the set of nonlinear, coupled integral-differential equations that support the eleven-dimensional framework for M-Theory,

"It is possible that Professor Steinmann is on the threshold of spawning a new era in physics that promises to merge Quantum Mechanics and General Relativity into a comprehensive and rational marriage for physics via M-Theory in the paradigm of string theory."

The NAS President believes that David might also confirm M-theory by demonstrating that the property of particle spin exhibited by Fermions. Fermions are particles obeying Fermi-Dirac Statistics, and Bosons, are energy quanta obeying Bose-Einstein Statistics, both fundamental to the well-accepted Standard Model for Particle Physics or SMPP. These particle properties are required as a basic, necessary condition for M-Theory. Half-integral spin, or Fermions and integral spin, or Bosons are the mathematical consequence of the basic assumptions of M-Theory. In physics, it is well accepted both theoretically and experimentally that this intrinsic property exists and is measurable, but its true basis for occurrence is not known. David believes that M-Theory requires that particle spin properties are an intrinsic and essential property for these two independent classes of fundamental particles.

The NAS President then closes his remarks by again welcoming all the guests to the final Grand Council Closing Conference and provocatively challenging David to demonstrate the existence of higher physical dimensions claimed by M-Theory.

David's presentation before the AAAS Meeting

David delivers a lucid and cogent presentation outlining the potential for M-Theory to be confirmed experimentally. As David concludes his remarks, he receives a resounding ovation from the audience. Afterwards, there is extended discussion and interchange during the subsequent "question and answer" session. Even subdued, adversarial comments acknowledge David's brilliant presentation, but doubts remain among many conference attendees, principally the old guard physicists, regarding the bold claims offered by M-Theory. These critics remain skeptical that additional dimensions other than the standard, four dimensional. spacetime model in General Relativity actually exist. These spacetime dimensions asserted by Einstein have been experimentally demonstrated and firmly confirmed by physicists. Experimentally inclined attendees are also skeptical that esoteric, mathematical models and obscure theoretical machinations are adequate to confirmed higher dimensionality proposed by M-Theory. Only unequivocal, independently confirmed experimental evidence that demonstrates additional dimensions is acceptable to this dominant and highly critical group. Thus, the Large Hadron Collider Experiments must confirm David's claims if M-Theory is to become a credible pursuit for mainstream physics.

David engages in vigorous debate with these critics, and the serial discussion only ends when Rebecca forcefully stands to gain David's attention. She dramatically, but tacitly, points to her Casio wristwatch, reminding David of her impending work schedule. David reluctantly informs the audience that his caretaker and transportation provider has declared an end to the discussion. David expresses appreciation for the attention of the audience and quickly ends the discussion. Rebecca demands that they leave now for her Emergency Ward shift at the Massachusetts General Hospital in Boston. David and Rebecca hastily gather all David's materials, and he and Rebecca begin their delayed and now rushed return trip to Boston.

David and Rebecca's rapid return to Boston (1 Oct Friday)

During the AAAS Conference while David was delivering his lecture at the Conference, unusually heavy, freezing rains have fallen along the eastern coast of New England; and the I-90 Interstate East from Albany to Boston is wet and slippery. Ominously, because their departure is delayed, Rebecca exceeds the safe limits for driving given the steadily worsening road conditions for the return trip.

The wind driven rain covers the Interstate in patches, and Rebecca's vehicle occasionally hydroplanes while travelling at speeds exceeding prudent travel constraints for the prevailing weather. David occupies his usual position in the passenger seat of the car and is exhausted from his presentation. He nods frequently, oblivious to the weather, as Rebecca concentrates on the highway traffic. She hazardously passes the slower vehicles travelling more cautiously in her path. As usual, David and Rebecca travel without conversation. David drifts in and out of slumber while Rebecca is intensely preoccupied with attempting to safely navigate her speeding vehicle in this hazardous environment.

Frustrated with the slower traffic, Rebecca complains to an inattentive David that the final Conference was protracted, and she must now drive perilously to reach Boston in time for her nursing shift. However, David is silent as usual and does not acknowledge or respond to her complaints.

Rebecca respects David's genius; but his singular, consuming passion for physics and his insensitivity for her work schedule often results in undue pressures upon her such as this hazardous and delayed return to Boston. David lives in his own world of science, and Rebecca leads a dull and monotonous existence of shift work and demanding nursing schedules at the hospital. Her medical duties are typical for an overworked and underpaid nurse at one of the largest and busiest hospitals in the world. Still, she recognizes it is David's career and his position that has liberated her from her mother's anger and vengeance for marrying a gentile, indeed a Muslim, an enemy of the Jews and Jewish State of Israel. Furthermore, David has provided Rebecca with financial security and safe refuge in America.

Tragic auto accident (1 Oct Friday)

The I-90 Interstate east roadway is now slippery and laden with pooled frozen water patches as Rebecca continues to compel the car's movement past the slower, cautiously moving traffic. As Rebecca rapidly approaches the I-95 interchange on the I-90 Interstate, a pickup truck in the left lane immediately in front of Rebecca suddenly crosses in front of her to make a last moment exit onto the I-95 Interchange north. With the rapid and violent change in direction of the pickup truck, a child's bicycle catapults from the pickup's bed and bounces onto the highway directly into the path of Rebecca's speeding car.

Rebecca panics at this unanticipated obstruction confronting her and she desperately swerves to avoid striking the rapidly approaching bicycle. She overcorrects the car's forward path and the car skids violently on the wet pavement, rotates from Rebecca's erratic steering and then overturns onto the car's roof. The vehicle, now on its flattened roof, careens and spins uncontrollably across the wet road and slams into the concrete abutment separating the Interstate highway and the exit ramp. The overturned vehicle is crushed by the ramp wall; and the flatten car roof and frame collapses, jamming the car's doors and imprisoning both of its occupants.

Rebecca and David are critically injured by the violent collision with the ramp wall. They are both wearing seat belts and the front passenger air bags rapidly deploy as the vehicle strikes the abutment. However, the crushing force of the impact against the abutment steel rail collapses the doorframe on David's side, and glass shards from the broken windshield and passenger side windows puncture Rebecca's air bag that immediately collapses providing her no protection from physical contact with the car's steering post. Both passengers are rendered unconscious from the violence of the accident. Rebecca is mortally injured. She is bleeding profusely from lacerations sustained from the glass fragments that have ruptured her airbag, pierced her face and lacerated and penetrated her torso and arms.

David and Rebecca are entombed, upside down in their collapsed, steel coffin. Fortunately, their crushed vehicle rests in the empty emergency lane on I-90, and no other vehicles collide with the upset Honda sedan. Passing vehicles slow as they come upon the crash, and cell telephones from these vehicle passengers quickly report the accident on the emergency number, 911.

Police and fire department personnel and equipment respond rapidly to the scene alerted by the emergency calls from passing cars. Rebecca and David have both sustained potentially fatal injuries. Rebecca continues to hemorrhage from multiple venous and arterial wounds to her head, neck and torso. Contusions, bone fractures and shock add to Rebecca's physical trauma from the automobile's steering post that has been thrust into passenger compartment by the impact of the vehicle with the Interstate barrier. David is protected by the passenger side airbag, but the car's passenger side door and frame have crushed in against his right hip and captive leg resulting in multiple fractures and contusions. However, David is not hemorrhaging from any major lacerations or wounds. His collateral injuries are bone fractures and contusions and are not immediately life threatening.

The Massachusetts Highway Patrol partially closes Interstate-90 in the vicinity surrounding the accident, and emergency personnel and EMTs in the fire engine dispatched to the scene quickly begin the difficult task of extracting the occupants within the severely damaged vehicle. Emergency personnel quickly recognize that Rebecca's severe hemorrhaging is life threatening and they focus their immediate activities on extracting Rebecca first from the upset vehicle. Her driver side door is forced open with wrecking bars, and the EMTs cut her seat belt, carefully lower, and remove her from the vehicle. They strap her bloody and battered body onto a stretcher and rapidly transport her to the Emergency Ward at Massachusetts General Hospital in Boston.

Because the deformed vehicle frame and doorpost imprisoning David are entrained with the road abutment, the emergency crew must now drag the overturned vehicle away from the abutment and cut through and remove the bent steel center post frame with metal saws to extricate David. Cutting through the doorframe is slow and prolonged to avoid additional injury to David. Furthermore, there is risk of fire from the sparks generated by the metal saws since gasoline is slowly seeping from the overturned vehicle despite the efforts of the fire engine crew to disperse the flammable fuel. Fortunately, the heavy rain has greatly reduced the risk of fire at the scene but exacerbates the rescue efforts of emergency personnel at the scene.

David's NDE (1 Oct Friday)

While entombed in the crushed, overturned vehicle and amid these myriad emergency rescue efforts, David experiences a "Near Death Experience" (NDE). This unique, subliminal, out-of-body experience will profoundly transform his life and the lives of many others, known and unknown to David.

As his NDE unfolds, David is aware that he is separated and detached from his physical body that is trapped in the upset vehicle. He witnesses, dispassionately, the emergency personnel frantically working to extract him from the twisted steel chassis. Although he observes the personnel and their activities with full comprehension, the emergency personnel are unaware of David's undetectable NDE presence. Remarkably, as David witnesses this scene of injury and potential fatality, he realizes he is without physical pain, fear or even emotional attachment to his unconscious physical body in this subliminal state; even though he recognizes his own severely injured body trapped within the wrecked vehicle.

David closes in towards his unconscious body. His injured face covered by his hands is pressed against the inflated air bag, and his arms are pressed against his chest with his hands covering his face in an apparent, disparate act to protect his head from injury. David also witnesses that his severely injured and bloodied right leg is imprisoned between the passenger seat and the bent doorframe.

David hears the irate words of the firefighter operating the metal saw that encounters resistance while cutting through the passenger-side steel doorpost. The loud, high-pitched blare of the saw blade crescendos over the clamor and flashing emergency lights that illuminate the accident scene. The metal sparks thrown by the saw shower David's motionless body. As the crewmembers continue their difficult rescue effort, David realizes he can understand not only their spoken words but comprehend their non-verbalized thoughts and emotions. Although David feels no stress or fear watching the emergency crew, he witnesses the men's strain and anxiety as they attempt their rescue of this entrapped passenger in the dark, wet and hostile weather surrounding the wreckage. David is profoundly aware that he can observe these activities and comprehend both the physical and mental intensions of all personnel at the scene of the accident even though he is unaffected and without pain or fear regarding the state of his injured body and the harsh environmental conditions.

David, in this NDE, finds he is positioned in space above the emergency personnel. He is aloft from accident scene without constraints of gravity for his position as he witnesses these events. Furthermore, the usual physical effects of exposure to the cold temperature, the wet weather, the glaring emergency lights and turmoil at this accident scene pose no distraction or stress upon David as he witnesses the emergency actions. Despite clamor and cacophony at the accident scene, he is fully cognizant of all actions and thoughts of those in his presence. He can move freely, observe from any position, and witness any scene he wishes surrounding the accident.

David now deliberates on Rebecca who has been removed from the wrecked vehicle. He is then immediately situated within the ambulance as it rapidly transports Rebecca to the Emergency Ward at Massachusetts General Hospital. An EMT in the vehicle, a slender, redheaded, female technician, has typed Rebecca's A, Rh positive blood and affixed a transfusion set infusing her body with whole plasma through the main artery in her left arm. The other attendant, a young male, Hispanic technician, has intubated Rebecca to support her respiration through her severely lacerated throat and damaged esophagus. The blood pressure cuff on Rebecca's right arm indicates that her heart rate is faint and rapid, and her blood pressure is depressed from extensive blood loss. David probes the EMT's thoughts and actions and determines that the EMT's are skilled and efficient in their desperate efforts to sustain her fragile hold on life. David also realizes that both EMT's believe Rebecca's injuries are so severe that she may succumb before the ambulance reaches the hospital.

When the ambulance arrives at the emergency entrance of the Hospital, Rebecca is immediately removed from the emergency vehicle and transported to the Emergency Ward on her stretcher. There an emergency triage team places her unconscious and bleeding body on a surgical table where an additional emergency trauma team immediately descends upon her applying a multiplicity of monitoring instruments and life support devices.

Rebecca dies (1 Oct Friday)

The Emergency Ward staff assesses the readings on the monitoring instruments and continues the blood transfusion and intubation with oxygen. Rebecca is now in full cardiac arrest and her own lungs have collapsed. The physicians and nurses in the Emergency Ward desperately attempt to restore her cardiac and respiratory function, but her massive loss of blood and severe trauma from the glass shards in her upper torso thwart the efforts of the staff. It is apparent that Rebecca's injuries are fatal and finally the Senior Emergency Staff physician declares that the team cannot revive her.

David then observes that Rebecca's spirit slowly ascends from her broken body. Without words, David confronts Rebecca's spirit that informs David that she is mortally injured with irreparable damage to her physical body, severe blood loss, and a brain impaired from anoxia. Rebecca has rejected her physical body and she will not remain in mortality. David implores her to return to her body. However, Rebecca again informs David that her massive loss of blood and protracted oxygen deprivation of her mind are irreversible. She refuses David's petition, and her spirit recedes from David's presence. He realizes that Rebecca is now physically dead, and only her lifeless, physical corpse lies on the emergency table at Massachusetts General Hospital. David observes the time on the digital clock in the Emergency Ward wall. The time is 11:31 PM, Eastern Standard Time when Rebecca is declared and recorded as officially deceased by the attending emergency staff physician.

David now wills his return to the accident scene to witness the continuing efforts of the emergency crew to extract his body from his tangled metal coffin. The emergency crew is at the same point in time when he

18

deliberated upon Rebecca's condition and he was positioned in the ambulance transferring her to Massachusetts General Hospital. David realizes that in his NDE state, he can transfer and position his presence in time and space through his simple volition.

David now wants to witness the earlier events leading to the automobile accident. He immediately observes Rebecca and him rapidly traveling in the car along the wet, icy Interstate I-90 Highway. Then as their vehicle approaches the north Interchange for I-95, an old Chevrolet pickup truck with Vermont license plates and a poorly secured load of household items in the truck bed suddenly swerves crossing both lanes in front of Rebecca. The sudden lane change by the pickup truck dislodges a child's small bicycle from the bed of the pickup. The bicycle bounces chaotically upon the pavement directly into the path of Rebecca's vehicle. David witnesses Rebecca's panic as she impulsively responds to this sudden obstruction with a desperate effort to avoid collision with the obstacle. However, her frantic actions shift the car's path and put the vehicle into an uncontrolled spin. The car then overturns, sliding on its roof across the wet pavement, and slams against the Interstate's concrete barrier. The vehicle smashes violently on the passenger side where David is buckled in his seat. Both air bags deploy. However, as the car impales against the exit barrier, the interstate barrier rail penetrates the windshield and the passenger door window. Glass shards from the windshield and side windows immediately puncture Rebecca's airbag as the vehicle collapses violently against the barrier. Rebecca's head and chest slam against the steering column that is torn from its mounting and driven forward impaling her body. The shards from the fractured glass lacerate Rebecca's face and torso. David's airbag protects him from the flying glass fragments and cushions the impact of his body against the car interior. He then observes the deforming passenger doorpost crush and imprison his right leg and fracture his hip and femur.

Again, as he witnesses these traumatic injuries to his physical body, David is without emotion or pain, but simply an aloof observer of the critical injuries sustained by the passengers within the upset vehicle. David's subliminal being is completely isolated from the trauma and panic of the situation. He has witnessed the accident scene with no emotional or physical perception of the injuries to his physical body. He is immune to the trauma wrought upon his physical body.

David returns to accident scene (1 Oct Friday)

David, still sequestered in his NDE, then returns to the wrecked vehicle that has now been dragged from the exit barrier to permit access to the passenger door. The emergency crew finally cuts through the side frame and removes the deformed passenger door. The crew then cuts David's lap belt, extracts his injured body from the twisted metal frame, and quickly places his unconscious form onto a stretcher. Two EMTs strap him to the stretcher, and David is inserted into the emergency vehicle for transport.

The emergency vehicle rapidly conveys him to Massachusetts General Hospital. David experiences the high-speed trip to the Emergency Ward with sirens and flashing lights attesting to David's serious injuries. Upon arrival at the Emergency Entrance, his broken and unconscious body is removed from the vehicle, and the body-laden stretcher is rushed into surgery by the emergency staff.

After an intense, but orderly medical assessment, the triage team leader declares that the male victim is seriously injured, but his injuries do not appear fatal; and he should survive. David witnesses an Emergency Ward nurse who states that the other victim of the crash, a young white female, the driver of the vehicle, has succumbed to her injuries. The emergency staff had not been able to save the earlier female crash victim, David's sister Rebecca. However, the EMT's rapid transport and the Hospital's effective medical intervention insures that this male victim should survive with a crushed right hip and upper thigh and surficial contusions, but limited blood loss. Extensive surgery will be required to repair the crushed hip and thigh; but his vital signs are stable and positive. Multiple surgical procedures will be required to correct his severe bone fractures and extensive muscle damage must wait until an accurate diagnostic assessment of the actual internal bone and tissue damage can be made.

David's NDE closes (1 Oct Friday)

David's NDE then closes as he experiences his gradual return to his physical body now located in the ICU where he will remain unconscious and oblivious to his physical pain and surroundings. Later, the next morning, after being moved into a private Hospital Suite on the fourth floor at Massachusetts General Hospital, in Boston, Massachusetts, he will awake.

CHAPTER 2 (2 Oct - 9 Oct)

David awakens at Massachusetts General Hospital (2 Oct Saturday)

As David slowly awakens, he observes that a traction bar immobilizes and supports his right leg that is tethered to an aluminum frame attached to a rigging supported by four corner posts at the side of his hospital bed. The throbbing pain from his right hip and leg is intense despite the slow drip of morphine being delivered into his venous blood stream by the intravenous lines draped near his bed. His left index finger is encapsulated with a blood gas monitor, and a network of intravenous probes and peripheral lines are connected to the monitoring equipment that continuously interrogates and records the biological state of his severely injured body.

David visually surveys his private Hospital Suite and observes the young female nurse administering to his medical care. She is a slim, statuesque, blonde-haired woman that David surmises is probably in her late twenties. Her golden blonde hair frames her angelic face that is accentuated by pale blue eyes and flawless, silken complexion. Her opulent blonde locks of hair are bundled with a tight band that is held with a blue silk ribbon that complements the hue of her blue eyes. As she moves, David observes that her supple legs are slender but strong like those of a ballerina that transport her slender form gracefully and precisely throughout David's suite. Despite his physical pain, he is fixated on the beautiful, young woman who sweeps gracefully and effortlessly through his suite as she administers to her patient.

David recognizes that she is diligently and skillfully maintaining the complex array of medical equipment and computational electronics that surrounds his hospital bed. She continually and carefully monitors and records his physical and biological conditions on the hospital suite computer located at the foot of his bed.

Then this nurse, the singular focus of his rapt attention, realizes that David is awake and quickly approaches his bedside and greets him respectfully and formally saying,

"Professor Steinmann, how do you feel? Is your pain very intense? I can increase the morphine delivery rate if your discomfort is too severe, Professor."

Christa Olsen is attending nurse (2 Oct Saturday)

Without verbally responding to her tender inquiry, David reads the nametag on her starched, white cotton blouse that displays her comely figure and small waist. In bold, black letters with the ubiquitous Massachusetts General Hospital insignia, the nametag reads,

"Christa Olsen RN (enfermero)."

David recognizes that the Spanish word enfermero denotes, "Registered Nurse."

Christa informs David that he has been unconscious for over eight hours from the heavy sedation he received in the Emergency Ward upon his arrival late last night. However, she is delighted and relieved to see he is now awake and responsive. She begins to recount David's medical management in the Emergency Ward since he arrived by ambulance to the hospital around midnight on Friday. David listens intently to the details of his medical treatment since his arrival at Massachusetts General Hospital from the automobile accident that took the life of his sister Rebecca. Christa then closes her factual, detailed medical soliloquy with the unofficial prognosis that with an extensive series of orthoscopic surgeries supported by intense physical therapy, his severely damaged right hip and leg can be corrected. He should fully recover from his serious injuries and be ambulatory. Then Christa smiles warmly and emphasizes to her intent patient,

"Professor David ben-Steinmann will then be able to resume his eminent position at Harvard University."

Christa then pauses, hesitates, and with obvious sympathy, informs David that his sister; Rebecca Steinmann, did not survive the tragic accident. Christa solemnly affirms this tragic event with subdued and poignant words,

"Dr. Steinmann, your sister, Rebecca, died in the Emergency Ward shortly after her arrival. The emergency staff, despite intensive intervention, was unable to save her life. She had sustained massive injuries and suffered severe blood loss and irreversible oxygen deprivation. I am very sorry to inform you of her death. It must be a great shock to you and your family to have lost your sister in that tragic accident."

David responds to Christa's compassionate narrative with the simple words,

"Thank you, Ms. Olsen, for informing me of her death and your empathy. I was aware she had succumbed from her injuries."

With no further verbal response from David, Christa then excuses herself to continue monitoring his medical parameters and record the data in his medical records using the Hospital suite's computer terminal. David continues his focused surveillance of his very attractive nurse as she continues her medical duties. After completing her medical monitoring data entries, Christa returns to David's bedside and resumes her effort to assess his psychological state after his life-threatening accident. She repeats her commiseration over the death of his sister and affirms her confidence in David's full recovery. Christa also informs David that the Hospital Administration has informed David's mother who lives in Israel of the tragic accident and Rebecca's death. Christa inquires if there are other family members who should be informed of his accident, the death of his sister, and David's present medical condition.

David nods his head conveying a negative response, but he remains tacit. Christa, still concerned that her patient may be suffering delayed shock from the trauma of the accident and his sister's death, searches for additional probing words of consolation and assessment of his mental state. As Christa continues to monitor his medical parameters, she smiles confidently and remarks,

"You should be able to resume your endowed chair at Harvard, Professor Steinmann. The entire staff at Massachusetts General Hospital is aware of the presence of our famous patient, the brilliant, young physicist who holds the Weisenhausern Endowed Chair in Physics at Harvard. Your presence at the Hospital is now known by the entire medical staff."

News media coverage of David (2 Oct Saturday)

Christa then remarks that national news reporters have been to the hospital interviewing the staff about David's physical condition. The media was anxious to assess the medical state of the International Leader in M-Theory. She has been part of the nursing staff serving him since his admission and will be his attending nurse this day. Christa continues her disclosure of earlier events and visitors in David's Hospital Suite declaring,

"Already you have had many visitors, including reporters from the Boston Globe, the New York Times, the Washington Post and earlier, a Time Magazine photographer who took pictures of you here in your suite while you were unconscious. I asked the photographer to wait until you were conscious, so we could gain your approval for his pictures; but he said he had a news deadline to meet and his editor demanded evidence of Professor Steinmann's presence at the hospital. So, I allowed him to photograph you. I hope you approve of my actions?"

David still does not respond to her inquiry but simply nods his head in approval of Christa's actions. Failing to gain verbal interaction from David to assess his emotional state, Christa then informs David that since he is now conscious and hopefully comfortable, there are important members of the medical staff who wish to speak with him. Christa excuses herself and leaves David's suite. She soon returns with a group of five men unknown to David. The men file into his suite, gather around his bed and individually greet him, expressing gratitude that he is now conscious and able to respond to their questions and medical assessments.

Hospital staff visits David (2 Oct Saturday)

The five men each identify themselves to David and declare their respective positions with the Massachusetts General Hospital and Harvard University. To David's surprise, the group who surrounds his bed in close quarters, includes the Dean of the Medical School at Harvard, the Senior Administrator of Massachusetts General Hospital, the Chief of Surgery and two attending senior resident orthopedic physicians.

The Dean of the Medical School informs David that the President of Harvard University would have joined them, but she is in Washington DC, testifying before a Joint House and Senate Congressional Committee regarding National Health Coverage Legislation. However, the Harvard President may contact David later upon her return from Washington.

David, witnessing this large group of medical inquisitors and their extensive and impressive medical credentials, quips to the imposing ensemble,

"Does the presence of such an eminent ensemble of medical and administrative VIP's mean I am dying?"

The group communally laughs, and the Chief of Surgery assures David that he will not allow David to succumb at Massachusetts General Hospital, at least not during his shift. The team members then converse among themselves, in precise but obscure medical terminology, regarding David's physical and medical diagnosis and preferred treatment options.

Then the Chief of Surgery briefly outlines to David, the potential series of reconstructive surgeries anticipated to restore his right hip and leg. One of the orthopedic surgeons states that CT scans confirm that David's right hip, femur, and associated muscles, tendons and nerves were severely injured in the automobile accident, and repair and restoration procedures will be numerous, intensive and challenging. However, the Chief of Surgery then counters that a successful series of progressive surgical procedures, absent of infection and combined with an intense prophylactic regimen of physical therapy, should result in David's full recovery and subsequent unrestricted ambulation. The Dean of the Medical School jests that David should be able to rejoin his long boat team in the Summer Regatta on the St. Charles River. The Dean then declares,

"The Massachusetts General Surgical Staff has delayed initiating this planned, intense and intricate series of surgical procedures until our eminent and most experienced orthopedic and plastic surgeon, Dr. Joseph Feinstein, returns from Israel. Dr. Feinstein specializes in such major reconstructive surgical protocols. The Harvard medical staff assessment is that the delay in immediate surgical intervention is warranted because the first surgical episode will be very invasive and exploratory and must be supervised by Dr. Feinstein. Dr. Feinstein can assess and confirm David's true medical state and injuries and he carefully establish the appropriate serial agenda for subsequent surgical protocols and recovery procedures."

David requests support for research (2 Oct Saturday)

The Senior Administrator for Massachusetts General then inquires what their famous young patient requires during his extended stay at Massachusetts General Hospital. The Administrator has already conferred with David's Physics Department Chair at Harvard. David's classes and postdoctoral seminars will be temporarily covered by other faculty members until David is able to resume his research and teaching duties at Harvard. The Massachusetts General Hospital and Harvard University enjoy a long association and successful cooperation in all areas of science including teaching, research and, of course, medicine. Massachusetts General Hospital is prepared to do whatever is necessary to make David's stay, medically successful, hopefully short, and allow him to resume his prominent work in physics. The Dean of the Harvard Medical School has been instructed by the President of Harvard University that Massachusetts General Hospital is to return, expeditiously, Harvard's latent Nobel Laureate to his eminent research in physics. The Dean then confides with the others in the group that Harvard is unduly lean with these prestigious appointments presently. The Harvard University international reputation and endowment needs scholarly enhancement and financial stimulus.

In response to the Hospital Administrator's offer of services to support him, David quickly asserts,

"I must continue my theoretical work in M-Theory. With the recent startup of the LHC or Large Hadron Collider at CERN, the acronym for the European Center for Nuclear Research in Switzerland, there are several experiments that I am expected to theoretically support and analyze to confirm certain, critical parameters in M-Theory. During my presentation at the AAAS Conference before that fateful accident that took the life of my sister, I had gained tentative support for these experiments. I must continue and confirm that support."

"My papers and personnel computer that contain this important work are at my apartment in Belmont. These materials must be retrieved, and efforts provided by the staff here at the Hospital to allow and support me to continue my work. I can still operate a laptop computer from my hospital bed and continue my work. You can remove that superfluous flat screen television from my suite to provide me with more space for my research materials. If the Hospital has a spare, empty bookcase, I wish to have the bookcase placed where the television presently resides to hold my books and notes. Furthermore, with the death of my sister who lived with me in Belmont, all her belongings must be gathered, and her personal affairs closed. I anticipate that my mother will soon arrive from Israel, and she will demand closure for the death of her daughter. My strict, orthodox mother will insist on the return of Rebecca's body to Israel for interment."

The Dean of the Harvard Medical School assures David that these requests can and will be honored. All necessary provisions to allow him to continue his research will be made. However, David must realize that the most important and immediate need is to restore his severely injured right leg and make him ambulatory and physically active. The other members of the group mutually affirm the Dean's words.

David requests Christa as his nurse (2 Oct Saturday)

David, then, surprisingly, informs the group that to support these numerous tasks and provide the necessary medical services he will need while a patient resident at Massachusetts General Hospital; he wants Christa Olsen, the attending RN presently in his suite, to be singularly assigned and exclusively dedicated to his hospital care during his hospital confinement. David strongly confirms this request, declaring,

"I believe that Nurse Christa Olsen can and will provide the special nursing services I need. I am comfortable with her presence and confident that she is fully capable to provide my essential medical needs and support me, so I can continue my scientific research. Gentlemen, will you grant me her exclusive nursing services?"

The Administrator of Massachusetts General Hospital, surprised at David's statement, responds that that such a request can be honored if Nurse Olsen agrees to his petition. The Administrator calls Christa to David's bedside, informs her of David's petition, and asks for her approval of this unusual request. Startled and perplexed, Christa responds respectfully to the Administrator,

"If Professor Steinmann believes I can support his medical recovery and assist him with his research, I will gladly accept his request."

The Administrator then tells Christa that she will be immediately relieved of all other duties at Massachusetts General Hospital. Her principal and sole responsibility will be the exclusive care and support of Professor Steinmann while he remains a patient at the Hospital.

Christa acknowledges the Administrator's assignment and reconfirms her commitment to David's care and support for his research needs. She then excuses herself from the group to obtain David's regimen of prescribed drugs for the day from the Hospital Pharmacy. After Christa departs from the suite, David assures himself of Christa's departure and then confides with the gathered men by declaring wistfully,

"Besides her nursing skills, she is a very attractive young woman; and I will enjoy her presence as well as her excellent medical care. Perhaps, gentlemen, I may recover sooner with her sustained presence and superb nursing skills."

The amused Dean acknowledges David's words and then turns to the other members of the all-male group and quips that obviously David's libido was not injured in the accident. The men mutually laugh and acknowledge this subtle male nuance. Then each member of the group wishes David a rapid recovery as the members exit David's suite.

Christa's nursing duties for David (2 Oct Saturday)

Christa soon returns to David's suite. She administers David's prescribed oral medicines and resumes performing her routine nursing duties. She changes his urine bag, checks the IV lines for patency, and records his vital signs and blood gases in the patient's nursing report on the Hospital Suite medical computer. David watches her with unspoken, but rapt attention. While intently observing her in action, he concludes that she is very graceful but efficient, effeminate but strong, and, of course, very, very attractive. As Christa performs her numerous nursing duties, David poetically muses in his analytical mind that this beautiful, young woman moves like a gentle, laminar fluid...gliding smoothly and effortlessly through his hospital suite...flowing around obstacles and...effectively and unobtrusively accomplishing her nursing tasks. In addition to her expert medical support and surveillance of David, she gathers and repositions articles out of place, dispenses clean and fresh items to replace the soiled ones, adjusts the furniture and gathers and replaces medical supplies and fixtures throughout the suite. In summary, this nurse, that he has successfully commandeered, completes all her duties for David's care expertly, gracefully, and without complaint. David is enthralled with Christa's presence, demeanor, and especially her physical beauty.

After intently observing her at work for over an hour, Christa becomes aware of David's continuous and rapt attention of her movements and duties. She attempts to ignore his intense, sustained observation of her routine nursing activities. She wishes to remain professional in her demeanor with her exclusive new patient, but Christa's curiosity has been aroused by David's request for her exclusive care of him.

Finally, David breaks his silence and inquires of Christa as she is recording his medical information on the computer at the foot of his bed,

"Ms. Olsen are you surprised and perhaps offended by my request for your exclusive and solitary assignment for my care?"

Christa pauses in her recording of medical data and approaches David and replies respectfully,

"I am surprised at your request Professor Steinmann; but if I can contribute to your recovery, then I am pleased to attend and serve such a distinguished patient. Your presence at the Hospital has been a major source of national news and excitement since you first arrived in the Emergency Ward."

Christa states that the entire hospital staff is aware that David ben-Steinmann, the Endowed Harvard Professor of Theoretical Physics, is a patient at Massachusetts General Hospital. Christa, with a provocative grin, then admits demurely, but with obvious satisfaction,

"Several of the other female nurses in the Hospital were vying for your nursing care. However, apparently, you believe I can adequately serve you. I am truly honored you have selected me, and I shall do my very best to serve you and meet your expectations for my medical service, Professor Steinmann."

David responds simply,

"Yes, Ms. Olsen, I will greatly appreciate your medical service and attendance."

David then inquiries about Christa's regular schedule for routine nursing care at Massachusetts General Hospital saying,

"When do you come on duty for me and what should I do to become a compliant patient?"

Christa hesitates, approaches his bedside and then informs David with the specification,

"Normally, I would provide your nursing services during the regular weekday, daytime nursing schedule. However, I have been informed by the Hospital Administrator that my schedule will be adjusted to meet any preferences you may have."

Christa waits for David's reply, but he is silent. In the absence of David's response to Christa's statement, Christa then suggests the following schedule:

"Professor Steinmann, nine AM to five PM each week day would be ideal for my own needs. I must travel from my apartment in Watertown to the Hospital, and this requires a transportation transfer in route. First, I board a Boston Metro bus in Watertown and then transfer to MBTA, the Massachusetts Bay Transport Authority subway. Those later transit times are less crowded than the early shift between 7:00 AM to 8:30 AM and the closing shift from 3:30 PM to 5:00 PM. However, I will gladly accommodate to your requirements. I can come on the regular shift or another schedule if you prefer. Of course, I will also be on-call for you any time in the event of an emergency or other special need you may have."

Christa's nursing schedule (2 Oct Saturday)

David, grateful for acceptance of his request for her nursing services and wishing to accommodate Christa's preferences and gain her favor, affirms,

"Ms. Olsen, do you mind if I call you Christa? The nine AM to five PM weekday schedule is fully satisfactory, although I wish it were seven days a week and the time extended during your standard shift. However, then you would become exhausted, and you might come to detest me as your patient. Furthermore, the Hospital might experience labor law violations for such extended service."

With her hands, akimbo to her hips, below her slender waist and her countenance displaying a provocative grin, Christa focuses her pale blue eyes intently upon David and playfully remarks,

"Professor Steinmann, it is very unlikely that I could ever detest you, and I do hope you will call me, Christa. I am not married, and Ms. Olsen is unnecessarily formal. Since you are now my solitary patient here at the Hospital and I am responsible for your medical recovery, I hope we can become informal as I serve you."

David responds candidly,

"Christa, I wish you would call me David. I am hopeful that we will become very good friends and informal in our conservations and interactions. I enthusiastically surrender myself entirely to your excellent and proficient, medical care."

Christa smiles and says simply,

"Good…very well…David."

David continues, attempting to remain discrete as he declares,

"I am very impressed with your nursing demeanor. I have observed that you serve my medical needs, both routine and unpleasant, with skill and without embarrassment, especially coping with those unpleasant, odious disposal bags. I hope that you will also support me in my research. It is essential that I continue my work despite this tragic accident. Yes, Christa, your very alluring presence and needed support will make my extensive and protracted medical care tolerable, perhaps even enjoyable."

Christa again smiles, fully displaying her angelic countenance, as she demurely counters,

"Dr. Steinmann, I thought dedicated physicists were single minded scientists who were occupied only with physics and singularly committed to their research."

David, somewhat embarrassed by his bold remark, counters quickly,

"Christa, even single minded physicists recognize unique female beauty and charm. But, seriously, I do need your support to continue my research in theoretical physics. I have all my scientific papers and research references at my residence in Belmont; and I am hopeful that you can assist me in securing these materials. I desperately need them for my research. Perhaps you can locate them for me in my apartment and bring these materials to me, so I can continue my work. I especially need my Apple Laptop Computer and my Multi-Terabyte external hard drive. My mathematical equations are stored on the hard drive. Unfortunately, my other computer that I had in my possession in my vehicle was apparently destroyed in the automobile accident. However, all my essential data files are backed up and are on the hard drive at my residence. Furthermore, I can access these files from virtual storage in my cloud data file."

Christa responds that she will locate David's apartment keys among his personal effects in storage and ensure that someone enters David's apartment and secures his computer and research materials next week.

Later, Christa serves David his noon and evening meals. She brings her own meals from the Hospital Cafeteria and sits beside him as they both consume their food. She enthusiastically engages him in conversation regarding his research in physics and particularly the rudiments of M-Theory for which David is the founder. David carefully outlines the basic concepts of M-Theory and is impressed with Christa's interest and the incisive questions she poses to him as he unfolds the unusual features of the eleven dimensions in M-Theory.

Thus, begins David's first full conscious day in the Hospital as he converses and becomes acquainted with his permanently assigned nurse, Christa Olsen, RN at Massachusetts General Hospital. Later in the day, as Christa completes her Saturday nursing shift with David, she informs him that she will return to his Hospital Suite on Monday morning at 9:00 AM as they have agreed. She will then begin their mutually established nursing schedule. Christa reaffirms her goal and the Hospital's objective to achieve the rapid and full recovery of their eminent patient. Christa prepares a note with her home telephone number, pins the note near his pillow and advises David to call her at home if he has questions or needs her. She then gathers her purse, approaches David's bed, gently clasps his hand and wishes him a restful Sunday. David, delighted with Christa's acceptance for his exclusive care, expresses his appreciation that she will serve as his personal nurse. He will anxiously await her return on Monday morning. They exchange sincere words of farewell, and Christa departs for her apartment for the weekend.

Christa's Personal Journal (3 Oct Sunday)

As is the habit of many faithful Mormons, Christa is committed to maintaining a personal Journal of the major events in her life and her perceptions of these events. Each Sunday evening, usually before retiring, Christa reviews the past week's activities from Monday morning through Sunday evening. She then records the significant events during the past week and provides her assessment of their importance and aspirations in her life.

Christa began maintaining a personal journal while she was a student at Brigham Young University where she pursued her nursing degree. Later, her journal recorded other events in her life such as the acceptance of a nursing appointment at Massachusetts General Hospital in

Boston, the admission and support of her mother as a late stage cancer patient in a terminal care center in Utah, and her activities and church-callings in the Cambridge Ward in Cambridge, Massachusetts. Now of compelling interest, Christa's Journal will include her experiences with her newly assigned medical patient, Professor David Steinmann admitted to Massachusetts General Hospital in Boston.

Significantly, Christa's Journal privately records her very personal and intimate feelings and her assessments of these events in her life. She periodically reviews her assessments recorded in her Journal to evaluate her behavior and her personal worthiness. Christa earnestly seeks to model and discipline her life, as she believes she should, as a devout Christian and a Disciple of Christ. She has a strong testimony of her Savior, Jesus Christ and is grateful for the blessings she believes that she has received and witnessed in her life. Christa also has a very close and personal relationship with her Heavenly Father. In her prayers, she addresses God, the Father, as she would a loving and caring paternal father. Her own mortal father, a career Naval Officer and F-18 Aircraft Pilot is deceased, so she converses with her Heavenly Father with intimacy and sincerity as she would with her mortal father. She believes and sincerely expects that her Heavenly Father will support her and advise her through the promptings of the Holy Ghost.

She now will begin to record and assess her activities at the hospital and her nursing duties and interactions with David ben-Steinmann. On each Sunday evening, her journal will display salient portions of her personal experiences, as Christa records them in her journal, for the previous week's events. The journal will provide the chronological record of the events and actions that will eventually bring her and David together as she is prompted by the Holy Ghost. With her pen and personal journal in hand this Sunday evening, Christa ponders her new nursing assignment for David Steinmann and then records the following entry in her journal.

Christa's weekly Personal Journal for 27 Sep-3 Oct 2010

My permanently assigned patient for the duration of his commitment at Massachusetts General Hospital is David ben-Steinmann, the very young and brilliant, endowed Professor of Theoretical Physics at Harvard. He was seriously injured in an automobile accident that took the life of his sister. David's protracted medical treatment will require extensive surgical intervention and convalescence during his several months of confinement at Massachusetts General Hospital.

He has requested my exclusive nursing services during his extended stay in the hospital to address and correct his critical injuries. I am flattered with his choice, but uncertain of his motivation for selecting me.

He is a charming, handsome and provocative patient. I will follow his selection of me as his solitary nurse and the consequences of his choice with great interest. I feel prompted to faithfully record my association and experiences with David in my journal with due diligence. I believe I will be greatly blessed by this close and intensive nursing assignment with David ben-Steinmann. However, what the full consequences of his selection for me in my life, only my Heavenly Father knows.

CHAPTER 3 (4 Oct-10Oct)

Christa relates Emergency room admission (4 Oct Monday)

On Monday morning as Christa begins her solitary nursing duties in David's Hospital Suite, he questions her regarding his admission and treatment following his automobile accident. Christa again recounts David's treatment in the Emergency Ward in precise detail.

"I was on duty during your sister's arrival and then your arrival later in the Emergency Ward. Again, David, I am sincerely empathetic over the loss of your sister, Rebecca. I realize to have lost your sister and your only sibling in such a tragic event is very agonizing for you and your family. Rebecca was admitted with very critical injuries she sustained in the accident. Her injuries were so severe and extensive that the staff soon recognized that her recovery was very doubtful. She had suffered massive loss of blood during the accident, and her brain had been deprived of oxygen from severe trauma to her esophagus and respiratory system for a sufficient time that she would have sustained permanent brain damage. Once again, I express my deep personal sorrow and that of the entire emergency staff over our inability to save Rebecca's life."

David then comments,

"I appreciate your sympathy and the emergency staff's commiseration for her death. However, my sister did not wish to live. She fully realized the fatal consequences of her extensive injuries."

David's comments and assessment perplex Christa, but she continues with the account of other events in the Emergency Ward including David's arrival and the successful efforts by the emergency staff to sustain his life.

As Christa continues her account of Rebecca's and David's Emergency Ward procedures; to her surprise, David augments and corrects her statements regarding many of the events that transpired in the

Emergency Ward both for David and his sister, Rebecca. During Christa's account of Rebecca's demise, David responds to her saying,

"I know that Rebecca died at 11:31 PM from multiple injuries and trauma including the attending physician's final diagnosis of cardiac arrest and acute respiratory failure."

Incredulous over David's declaration, Christa asks David to explain the source of his detailed knowledge regarding events associated with Rebecca's death.

"David, how are you aware of what transpired in the Emergency Ward? You were not even present when your sister died; and later when you were admitted, you were unconscious and anesthetized during your entire stay in the Emergency Ward."

David avoids disclosing his Near-Death Experience and simply asks Christa if she will confirm the events and times as he recounts them to her. David then provides his own detailed, serial itinerary of all the events he witnessed at the hospital during his NDE. To Christa's amazement, she realizes that David's description is remarkably accurate and detailed. Indeed, he recalls events that Christa has forgotten or did not even witness.

Christa then emphatically inquires of David again,

"David, how can you be aware of those events that you have recounted to me with such accuracy and detail? Has someone from the Emergency Ward staff informed you of these events or have you accessed and read the medical reports?"

David informs Christa of NDE (4 Oct Monday)

David pauses for several moments while admiring Christa's pale blue eyes that extol her cameo-like face and her golden blonde hair. David then concludes he will confide with Christa regarding his NDE and responds thoughtfully with circumspection,

"No, Christa. You may be suspicious and skeptical with my explanation, but during the aftermath of the automobile accident, I experienced what is designated as a Near Death Experience or NDE. I witnessed, outside my physical body, all the events that I have related to you. I realize that my explanation is unbelievable and even unscientific. But my NDE is factual and true, even if irrational and exceptional."

David continues his explanation,

"As you must know from reviewing my personal hospital records, I am a diagnosed and confirmed epileptic; so, I originally believed that my subconscious experience during the accident was, as I recall from the previous official medical description for an epileptic seizure, 'a transient disturbance of cerebral function due to an abnormal paroxysmal neuronal discharge in the brain.' However, the verifiable events surrounding the accident now make me believe it was a true and accurate NDE I experienced and not an epileptic seizure. Furthermore, a seizure would not provide me with such detailed and apparently accurate information and experiences that occurred in the absence of my physical or conscious presence."

Christa is without explanation for David's NDE. Drawing closer to his bedside and pondering his declaration, she remarks thoughtfully,

"I have heard and read of such subliminal episodes, but I have never been directly involved with a patient's NDE. However, your account is remarkably detailed and very convincing, David. I will undertake to confirm the accuracy of your account from the hospital records for those events and times you have stated. If you are correct, and I believe you are, then we must both explore and deliberate upon the meaning and significance of your NDE."

David, grateful for her understanding, responds,

"Christa, my NDE astonishes me, and it challenges my scientific training for rational assessment and explanation of any physical experience. However, in my subliminal NDE, I was able to perform remarkable acts and witness events that are not physically possible in our conscious universe. I was able to move through space without the constraints of gravity and the sensory effects of the physical environment surrounding my conscious being. Furthermore, I could also freely move forward and backward in time, observing past, present, and even future events. Christa, I could 'will my being' or spirit, if that were my subliminal state while I was detached from my body, to move and be at any location in space and at any time that I wished."

David relates his NDE to M-Theory (4 Oct Monday)

"Remarkably, my scientific research in M-Theory assumes the existence of more dimensions than our four spacetime dimensions and entertains the possible existence of conditions and capabilities that I experienced in my NDE. M-Theory contemplates the existence of a parallel universe surrounding our universe and separated by a membrane or barrier that divides these two states of existence. The parallel universe surrounding our present universe could possess far different properties in space, mass, energy and event time. Therefore, to have, rationally experienced such properties in my NDE, is challenging and very provocative for my research pursuits in M-Theory."

"Christa, I was outside and independent of my body during the NDE. I observed the severe physical injuries that my body sustained in the accident, but my extra sensual being or NDE spirit felt no pain or emotional attachment to that injured body. Also, the harsh weather and environmental conditions, the darkness, cold blustery winds, and heavy rain prevailing at the accident did not affect or even concern me. I was simply an independent, impartial observer of a tragic incident, such as you might causally watch regarding an accident scene on television and without any emotional involvement with the victims or their circumstances. But in this case, I was the severely injured victim in the accident"

David pauses, carefully pondering the possible impact of his NDE upon Christa, and then declares,

"Christa, this NDE, if confirmed by your inquiry regarding the facts, events, and times I have described may prove to be a major milestone in my pursuit of M-Theory. This NDE could provide me with independent, confirmed, personal evidence that our universe is surrounded by another universe, a parallel universe or multiverse with unique properties and different physical laws than our ordinary visible universe exhibits and permits. M-Theory contemplates the existence of another universe, incorporating dark matter and dark energy that co-existed with the 'big bang' origin of our current universe 13.8 billion years ago. This other, so-called dark, parallel universe comprises the missing 95 percent of the mass and energy hidden in our observable universe. Christa, the events I experienced in my NDE have profound implications for the physical properties of this virtual universe that may surround our own universe. "

Christa will confirm NDE (4 Oct Monday)

Christa listens intently as David continues relating the account of the other events at the accident scene associated with his NDE. As David concludes his words, Christa, with obvious excitement evident in her pale blue eyes and warm voice, then commits to David,

"I will confirm your account with other members of the Hospital Staff and the Hospital Records. You are correct David, if your account is confirmed; you may have penetrated the veil that separates our present world from the Kingdom of Heaven."

Christa now expresses her assessment of David's NDE disclosure with her personal testimony.

"David, I am a Mormon, a member of the Church of Jesus Christ of Latter Day Saints. We believe there is a veil or curtain, perhaps a membrane as you postulate in M-Theory, which separates our mortal world from the Kingdom of Heaven. The Kingdom of Heaven is the residence of God, our Heavenly Father and His Servants. We also believe that this veil is pierced or bridged when God bestows revelation through the Holy Ghost upon His

Prophets and other mortals whom God selects. It is possible that during your NDE, you became a brief visitor within God's Domain to observe what is ubiquitous to God. I am excited for you David. Your NDE may have provided you with evidence, perhaps even a personal revelation, of God's Domain. I believe that your NDE may be God's personal disclosure to inspire you and direct your research in science. I have diligently sought and read articles regarding M-Theory since you became my patient, and I now believe that God has plans for you and your research. I do hope you will continue your scientific investigations and discover the full meaning and consequences of your NDE."

David responds to Christa's affirmative declaration and avid statement of support,

"Thank you, Christa, for your confidence and support of my admission of my NDE. I was very concerned that you might suppose that my NDE was simply a hallucination resulting from the trauma of the automobile accident. Therefore, I am grateful for your trust in my claims. Perhaps we can pursue this unique experience together, since we will be spending considerable time with one another. However, please hold my disclosure confidential and discuss my NDE experience only with me. Others might disparage my experience and even ridicule me for my claims. There are many critics of M-Theory among my scientific colleagues. They would certainly deride and ridicule my admission of an NDE. I must carefully deliberate and evaluate the meaning and consequences of my NDE and its impact upon me and my investigations and research into M-Theory. Furthermore, I will share my findings only with you since you believe me and can support my investigations."

David then pauses and says to Christa,

"However, Christa, I am amused that you believe God, your Christian God, has plans or even interest in me, a scientist, indeed, a secular Jew."

Christa quickly responds to David's words by confidently declaring,

"David, my God, your God, the God of all mankind has great interest in both you and me and loves all of His spiritual children. As a Jew, you are a member of the House of Israel. David, you are a direct descendent of the Tribe of Judah through which the Son of God, Jesus Christ, who is the Messiah, came to Earth. Your NDE may be God's endorsement of your research efforts. I believe your NDE has and will greatly influence your life. Furthermore, I will be delighted and honored to pursue your NDE and disclose its full significance for you and perhaps even for me. David, I am excited over our mutual quest."

David smiles warmly at his beautiful nurse and softly answers,

"Thank you, Christa for your understanding and support. Our mutual pursuit may prove very eventful for both of us."

Medical Staff assess David's injuries (4 Oct Monday)

Later on Monday afternoon, Christa informs David that the Hospital Staff, headed by Dr. Joseph Feinstein will visit him at 2:00 PM to inform him of the assessment made by the Massachusetts General Hospital Orthopedic and Surgical Staff regarding his physical injuries. The Staff will present their proposed treatment plans to repair David's severely injured right hip, thigh and leg and restore his mobility and strength.

Before the Staff arrives in David's suite, Christa insures that David's medical dressings are freshly applied, and he is groomed and prepared for the Staff's physical examination of his injuries. She gently sponges David's face, arms, chest and right leg for the pending medical examination and replaces his top bed sheet. She then reorganizes the suite to accommodate the numerous Hospital Staff members that will surround his bedside during their examination. Completing her preparations for his examination, Christa pauses, smiles and confidently declares to David with obvious satisfaction,

"There, my handsome, young ward is prepared to confront the Hospital Staff and submit to their intrusive examination and diagnosis."

45

At 2:10 PM, the Orthopedic and Surgical Staff composed of Dr. Joseph Feinstein, two orthopedic surgeons, an anesthesiologist, and a surgical nurse enter David's Hospital Suite. The Staff members are greeted by Christa as they gather around David. David observes that Dr. Feinstein fits the classic image of a senior, erudite, Jewish physician. He is tall and slender-built with penetrating dark brown eyes, accentuated by gold-rimmed glasses and a well-trimmed beard and thin moustache. It is obvious to David that Dr. Feinstein is highly regarded at Massachusetts General Hospital and this accompanying staff exhibit great respect and deference for his position and vast medical knowledge.

After brief instructions to his companions, Dr. Feinstein asks Christa if the Hospital's Endowed Harvard Professor has been a compliant patient. Christa politely responds,

"Dr. Feinstein, as you are aware, I have been assigned to Professor Steinmann's exclusive care by the Hospital Administrator, and we are all committed to an expeditious return of this Eminent Professor of Theoretical Physics to his active teaching and research position at Harvard."

Dr. Feinstein affirms Christa's verbal assessment saying,

"Thank you, Nurse Olsen, we all acknowledge that he has received excellent care from you and the other members of the Nursing Staff."

Then turning to David, Dr. Feinstein says in Hebrew,

"Shalom" (Good afternoon), "ma shlomkha?" (How are you?)

David briefly responds in Hebrew.

"B'seder." (Okay)

Dr. Feinstein then continues his verbal exchange with the medical staff and David in English saying,

"David, I have just returned from Israel on Sunday evening. I was informed, while in Jerusalem, of the serious automobile accident that claimed the life of your sister, Rebecca Steinmann, and severely injured you. The Hospital Administrator emailed me in my room at the David Citadel Hotel and asked me to contact your Mother, Sariah Steinmann, and inform her of the death of her daughter and your serious injuries. The Hotel telephone operator acquired the telephone number and connected me with your Mother. I reluctantly informed her of your tragic automobile accident. Of course, she was devastated over her loss and informed me that she would immediately arrange for her travel to the United States. The Israeli Airline, El Al, has informed me she will arrive tomorrow, Tuesday morning, to visit you and reclaim the body of her daughter Rebecca for return to Israel for interment. I was deeply saddened to deliver such tragic news to your mother, but the Massachusetts General Hospital Staff realized you were unable to inform your mother of the tragic accident, so I did."

David responds to Dr. Feinstein by expressing his appreciation for his difficult, but essential task.

Christa then politely intervenes and says to the gathered medical team,

"Doctors, before you begin your close physical examination of my patient, I wish to respectfully caution this large medical team that strict hygienic protocol must be observed to avoid possible transmission of pathogens to my patient. He has no allergic reactions to Latex, so I have brought an ample supply of Latex examination gloves for your use if you need to physically contact my patient."

Dr. Feinstein affirms Christa's cautionary concern with an approving nod while all the attendees don examination gloves. Then Dr. Feinstein declares to David,

"Now allow me and my medical colleagues examine you and then inform you of the series of surgical procedures we propose to repair your fractured pelvic girdle and your compound, splintered femur. The pelvic girdle break is a simple, line fracture as shown in your CT scans and should be easily repaired. But the femur and surrounding muscles and tissues will probably require several invasive, serial surgeries to restore the femur's structural competence, insure adequate revascularization and bone marrow function, reattach and repair the damaged tendons and muscles, and hopefully restore nerve function for tactile perception and adequate ambulatory function."

Medical team assessment

The medical team members, led by Dr. Feinstein and supported by Christa, then perform a methodical and comprehensive examination of David. After considerable probing, interaction and consultation, the medical team completes their examination with a detailed review and assessment of the medical records and numerous CT scans that were taken since David's arrival at the Hospital.

Concluding the examination, Dr. Feinstein informs David that the Hospital Staff proposes three serial surgeries each separated by a four-week recuperation and stabilization period. If these surgical procedures prove successful, three serial surgeries, combined with intensive physical therapy should be sufficient to restore the skeletal, biological, and neurological competence of David's right leg. He should then be ambulatory and without major residual impairment. Restoration of the full agility and strength in his right leg will require several months of continued physical therapy and cautious surveillance of his activities. Dr. Feinstein then closes his diagnosis and prognosis saying,

"David, do you agree with our assessment and the surgical protocols planned for the medical treatment of your injured right leg?"

David carefully reflects on Dr. Feinstein assessment and then responds,

"Dr. Feinstein, my experience and competence in such medical matters is very limited, and I defer to you and my nurse, Ms. Olsen, for advice and consent. Christa, what do you think of this assessment, well it be successful, so you can return to other, less demanding and more compliant and affable patients?"

Christa smiles warmly with affection and responds impishly while gazing ardently upon David,

"Yes, Professor Steinmann, this extensive and critical protocol should prove successful and be adequate to remove you from my duty slate, discharge you from this Hospital, and return you to Harvard."

Dr. Feinstein is amused at Christa's playful response, recognizing Christa's intended "double entendre". He then voices a closing "Shalom" to David as he departs David's suite accompanied by the other members of the medical team.

After the team departs, Christa moves close to David's bedside and softly and tenderly declares,

"David, you will not be able to liberate yourself from my nursing care that easily. I am here for the duration of your recovery. My dear Professor, I want you ambulatory again and able to leave this Hospital walking."

David gazes intently at Christa's large, appealing, pale blue eyes that accentuate her flawless complexion and the band of golden blond hair that is bound with a blue ribbon that drapes over her slim shoulders. David slowly recovers from his visual examination of his stunningly beautiful nurse and then declares poignantly,

"Thank you, Christa, with you providing my nursing care, we'll mend this sparrow's broken limb; and I will leave this Hospital walking sprightly. Perhaps, you will even teach me to dance after my release, which I cannot do now and have never done with any partner, beautiful or otherwise."

Smiling, Christa responds enticingly,

"It is a date, Dr. Steinmann; and I will not let you forget that commitment. I love to dance."

David's mother Sariah arrives at Hospital (5 Oct Tuesday)

On Tuesday morning, as David was informed by Dr. Feinstein, his mother, Sariah Steinmann arrives at Boston's Logan Airport following a twelve-hour, nonstop flight from Ben-Gurion Airport in Israel. After clearing

US Customs, she leaves her luggage in storage at the airport, exits the terminal, hails a taxi in the passenger-loading zone and immediately travels to Massachusetts General Hospital. Clutched in her hand is the twisted Jerusalem Post account of the tragic accident that severely injured Harvard's Endowed, Jewish Professor of Physics who leads M-Theory research.

She quickly exits her taxi paying only the metered fare without tipping. She curses the unsuspecting driver in Yiddish for his delays in Boston traffic and then enters the Massachusetts General Hospital Main Entrance. She demands the location of David ben-Steinmann's Medical Suite at the information desk. Taking the elevator to the fourth floor, she quickly exists and marches down the corridor to Private Suite 4110 where the room sign reads "David Steinmann", enter only upon invitation. Ignoring the entrance restriction, she bolts into David's private suite. She proceeds directly to David's bed and begins to verbally assault him in Yiddish (English translation follows).

Sariah confronts David (5 Oct Tuesday)

"David ben-Steinmann, you are responsible for 'killing' Rebecca. If you had not allowed her to join you in America and then forced her to drive you to Albany, New York in such dangerous weather, she would still be alive. Your actions are unforgiveable and reprehensible. I loathe you for your recklessness. I have lost my husband and now my daughter to the tragic and unwarranted actions of others. You should not have allowed her to live with you in this overcrowded, Gentile City and left her with me in Israel."

David counters his mother's accusations in Yiddish and reminds her that she destroyed Rebecca's marriage by initiating a contested annulment. That was the reason Rebecca joined David in America. He provided a safe refuge for Rebecca to escape from her mother's domination. With this verbal rebuttal by David, Sariah strikes David sharply across the face and curses him again.

"David ben-Steinmann, I hope you remain a 'cripple' for your offenses against me and Rebecca. I will return Rebecca to Zion and inter my daughter where she belongs in Jerusalem. She will not remain in this cursed, heathen land of Gentiles that took the life of my husband and now Rebecca."

Sariah concludes her oral tirade,

"Americans are more accursed and anti-Semitic than the Arabs and their terrorist bands. At least Arabs are the expelled offspring of Abraham."

Christa is silently attending her duties in David's suite and is unaware of the meaning of the tense, harsh Yiddish verbal exchange between the mother and son. However, when Christa witnesses Sariah's physical assault on David, she rushes to David's bedside and attempts to intervene and restrain Sariah from further attacks upon David.

Sariah struggles with Christa, denouncing Christa in Yiddish as a blond, damnable "Christian Gentile."

Unable to restrain Sariah, Christa quickly exits the suite and secures two male hospital attendants who return and bodily remove Sariah from David's suite. Still cursing David, Christa and the attendants, Sariah is escorted from the hospital.

Confirming Sariah's departure from the Hospital, Christa then returns to David's suite and quickly examines his medical probes and monitoring lines to ensure their secure application and proper operation. Christa is visibly upset over this incident and replies to David with remorse,

"David, I apologize for allowing your mother to assault you. You told me you expected her to arrive today to return Rebecca to Israel. But I did not realize the extent of her anger at you over the death of her daughter."

David explains Sariah's anger (5 Oct Tuesday)

David, however, apologizes for his mother's behavior and tells Christa that he anticipated this would occur when his mother confronted him. Then David ruefully laments,

"She has lost her daughter, and I did contribute to the events that resulted in Rebecca's death. Had I not delayed my stay in Albany at the conference, Rebecca might not have lost her life in the tragic accident that was caused by her compulsion to return rapidly for her night shift at Massachusetts General Hospital. I suffer continued guilt and remorse for Rebecca's death."

David then briefly translates the verbal exchange in Yiddish with his mother that occurred. Christa now realizes the full extent of the guilt and remorse that David still harbors, and Christa must assuage David's anguish over Rebecca's death. Christa inquiries about the source of his mother's enmity towards Gentiles and especially Americans. David recounts his childhood in Israel and the distorted reasons for his mother's hatred of Americans. David relates the events leading to the death of his Father in a joint Israeli-American clandestine, anti-terrorist mission. David's father was killed in the crossfire during a hostile exchange with PLO terrorists. Then David details his home life with his parents and his mother's dominance of all family members including Rebecca, David and even his father. Finally, David concludes his candid assessment of his mother's demeanor,

"My mother is a vindictive, angry woman, who perceives a hostile and evil world surrounding her and her Zionist people. A world that she believes has and continues to persecute her chosen people, the Jews. She is a rigid, rabbinical, orthodox Jewish woman who cannot forgive past offenses against the Jews and her perception of unmitigated anti-Semitism evidenced by all Gentiles."

Then David pauses and somberly declares,

"But in her defense, Christa, she has lost her daughter; and I am responsible for the tragic accident that took Rebecca's life."

Christa consoles David about Rebecca's death (5 Oct Tuesday)

Listening compassionately to David's account, Christa counters David's words attempting to assuage his obvious anguish and remorse over the death of his sister.

"David, it was a tragic accident that assailed you and your family. Rebecca was driving dangerously for the given weather conditions. However, the true cause of the accident was Rebecca's dangerous change in lane and the cargo dislodged from the pickup truck in Rebecca's path. Those events were outside your control and responsibility. Finally, Rebecca was the driver of the vehicle. It was her improper response to the dislodged cargo that led to the tragic accident. Our lives are filled with events and the actions of others that we cannot anticipate or control their consequences. Rebecca realized that the injuries she sustained were fatal as you witnessed in your NDE associated with the accident. But, you will be with her again, together and reunited with your father and all your other deceased relatives. I believe that all humankind will be resurrected through the universal Atonement of Jesus Christ. All the tragedies, adversities and evils in this mortal world will be absolved and rectified by the Savior's universal Atonement. This I truly know and testify to you of the reality of Christ's Atonement. David; I hope that you will come to understand this truth and believe this fact also."

David gazes intently upon Christa's compassionate and appealing image and thoughtfully utters,

"Perhaps that will occur, Christa, I truly hope so. You are a very persuasive and choice woman. Besides your physical beauty and gifted intelligence, you possess immense wisdom and unusual compassion for one so young."

Christa confirms David's NDE (5 Oct Tuesday)

Christa has investigated the times and events David reported to her in his NDE with hospital records and other personnel in the Emergency Ward staff when David and Rebecca were admitted. Christa informs David of the amazing accuracy and detail of his NDE observations at the hospital Emergency Ward. They discuss at length the significant of his NDE observations. Christa's confirmation of these events perplexes and amazes Christa and excites David. Later, additional evidence will further confirm the accuracy and precision of the all the events David witnessed and reported in his NDE, including those at the accident scene, in the emergency room, and in the ambulances that transported Rebecca and then David to the Hospital. .

David is now intensely motivated by the significance of his NDE and especially the physical circumstances surrounding his own state of being during the NDE events. Remarkable, during his NDE, is the fact that David was in a spatial and temporal realm external to the physical state occupied by the people and events he witnessed. Also, the people he observed were unaware of his David's presence and external to his being and presence; but David was fully cognizant of both the actions and thoughts that transpired.

Significantly, In David's subliminal state, he was not subject to gravity and the prevailing physical environment. His own expected human discomfort due to heat and cold, fear and anger, pain and anguish that was present at the accident scene were absent in his NDE. He was not affected by the usual senses and feelings that a human would exhibit under such hostile and stressful conditions. His state of being was that of an aloof, objective, omnipresent observer. David was, in this setting, an intimate, but unbiased spectator observing his surroundings devoid of judgment or pain and without being controlled or threatened by the conditions he observed. He could objectively witness and perceive all that was occurring in his presence and understand spoken and even mental communications between individuals and their own personal thoughts and intentions. David's

NDE was an incredible, unique experience and foreign to all his past conscious experiences.

Furthermore, in his NDE, David could move forward and backward in time, and he was able to travel instantly through space by simply wishing to do so. What were the ramifications and consequences of these extraordinary capabilities in his NDE? Had he witnessed and occupied additional spatial dimensions where travel in time both forward and backward was possible? Incredibly, had he experienced the extension of dimensions in space and movement through time that he was seeking to confirm scientifically by M-Theory? What did his NDE portend for M-Theory? Was his NDE a confirmation of a parallel universe encompassing our present universe? Were Christa's beliefs in the Kingdom of Heaven true? Would his extraordinary experience with this NDE, his present circumstance as a patient at Massachusetts General Hospital, and the dominant presence of Christa be a prelude to major alterations in David's life and his research in M-Theory? David concludes that he must explore and evaluate these events with great care and evaluation.

First Surgery for David's injury (6 Oct Wednesday)

David's fractured pelvic girdle and broken femur pose a serious medical challenge if he is to walk again without serious impairment. The Chief Orthopedic surgeon at Massachusetts General Hospital, Dr. Feinstein, has meticulously examined detailed CT scans and proposed an extensive series of surgical procedures to restore the anatomical configuration and competence of his right hip and femur. However, the complications that may arise during his medical treatment are serious and unknown. The femur is an important source for producing erythrocytes and platelets for the body. Extensive, uncorrected damage to the femur could compromise this essential metabolic function as well as impair his mobility. Can the extensive set of serial surgeries properly repair and restore the strength and function of his shattered hip and leg bones? Will he be able to walk and perform with his previous agility and stamina? Fortunately, his left hip and femur are undamaged and functioning properly, and he should retain adequate hematopoiesis to support his biological needs. However, careful monitoring of his blood cells and hematocrit will be essential during these surgical procedures to restore the competence of his damaged femur. In addition, there is the added complication associated with David's epilepsy that may be exacerbated by the trauma from multiple surgeries and severe medical intervention.

The first major surgery that David will undergo is directed at realigning and rejoining the skeletal bones in his hip and femur into their proper position to begin the slow process of mending and restoration of structural competence of his skeletal frame. This surgery must be performed precisely and correctly to restore and maintain adequate blood flow throughout the bones and muscles of the right leg, otherwise permanent damage and paralysis could result.

David is scheduled for his first surgical procedure at 9:00 AM Wednesday morning. Christa arrives early at 6:00 AM on Wednesday morning to prepare her patient for surgery. She is intensely tasked with supporting the pre-surgical staff as they extensively examine David, review medical charts and CT scans, and assess his tolerance of anesthesia agents and other surgical and postsurgical drugs. Another confirmatory series of CT scans of his right pelvis and femur are performed, and David is then given a sponge bath by a male orderly in preparation for surgery.

Christa's medical record for David is reviewed by the staff and scrutinized by Dr. Feinstein. The anesthesiologist makes a short visit before David is taken into the Surgical Suite to reconfirm the proper anesthetic regimen and avoid possible inception of David's incipient epilepsy. After David is moved from his bed to the surgical table, Christa grasps David's right hand with both of her hands. Then with an affectionate, but anxious smile she declares soberly,

"David, I will pray for you during your surgery. I want you to recover rapidly and return to your Hospital Suite and to me and my care."

David responds warmly to her entreaty and vows tenderly,

"Thank you, Christa. When I return, I want to see your beautiful countenance first upon awakening after surgery. Please be here when I regain consciousness."

Christa responds with the simple, but poignant response,

"I will be here for you, David."

The surgical orderlies then transport David to Surgery on the sixth floor of Massachusetts General Hospital for his first major surgical intervention.

David's surgical recovery (6 Oct Wednesday)

After the prolonged, intensive six-hour surgical episode on Wednesday, David, still anesthetized, is taken to post-surgical recovery suite where Christa anxiously awaits the return of her unconscious patient. While in recovery with Christa fully engaged in monitoring David's biological state, she is visited by Dr. Feinstein who informs her that the surgery was very successful and without complications. His prognosis is that David should be able to leave the Hospital in early January, walking with his right hip and leg structurally competent and fully functional. Dr. Feinstein observes the marked relief on Christa's face with his prognosis. Then he gently probes Christa with a paternal inquiry.

"Christa, I have observed that you are very concerned and attentive for our young patient and, perhaps, even very fond of him."

Christa smiles at Dr. Feinstein and responds with sincerity,

"Yes, Doctor, you are very perceptive, and I do care for this young Jewish Scholar a great deal. I want him to recovery quickly and completely. He is a very important person in science, and he must return to his important research."

Dr. Feinstein muses over Christa's words and then adds parenthetically with fatherly sentiment,

"And I believe David must also return to you and your attentive nursing care, Christa. I hope you and David can develop that strong, intimate relationship for each another that I perceive is slowly incubating during his hospital admission. I have observed how well both of you interact together and the great concern and care you exercise for our injured patient. Yes, Christa, my dear young woman, a young, brilliant and agnostic Jewish male scientist and a devout, beautiful female, Christian Gentile, indeed a Mormon...a unique and improbable union indeed. I will wait to observe the outcome of this embryonic union with curiosity and anticipation."

Dr. Feinstein leaves the recovery room with Christa still gazing intently upon her unconscious patient. As she ponders Dr. Feinstein's words while observing the heavily sedated, unconscious David, she realizes that David is the man that she could and would marry if the circumstances were correct. Her feelings for this patient, who is now her singular responsibility, have dominated her thoughts and interests since her assignment for his care. However, David is Jewish and a brilliant and famous scientist at Harvard. She is simply a nurse, an unmarried, Christian, Mormon nurse at Massachusetts General Hospital. Would David consider marriage to her and accept the Gospel of Jesus Christ? She could only marry a man who could eventually take her to the Temple. Remarkably, David has chosen her for his solitary nursing care, and their mutual feelings for one another are slowly, but progressively intensifying. Is their fertile relationship prompted by the Holy Ghost? Can she proselyte and convert David, this eminent Jewish scientist, to Mormonism and have him baptized and confirmed a member of the Church of Jesus Christ of Latter Day Saints? If she could surmount all these immense obstacles, then David would be her chosen companion.

After David's return to his own suite on Friday morning from post-surgical recovery, he is still unconscious from the anesthesia administered during his extended surgery and the pain suppressants. Christa is intensely tasked with performing the postsurgical outfitting and monitoring for her patient. Christa attaches the IV line that delivers Ringers Lactate with the specific antibiotics prescribed for pathogen control within the extensive surgical field that traversed from his right hip to his ankle. Christa then gently sponges David's face, neck and arms and applies the auxiliary monitoring instruments to measure his blood pressure, blood gases and other vital signs. She insures each of the monitoring instruments is functioning properly, that all postsurgical recovery measures are satisfactory, and no serious issues exist in his vital conditions. Then, while she is entering the postsurgical blood-gas data into the Hospital Suite computer, David stirs and repositions himself in the bed as he slowly regains consciousness. He rubs his eyes and then drowsily calls for Christa.

David awakes from surgery (7Oct Thursday)

"Christa, are you here? Christa, where are you?"

Christa quickly approaches David's bedside, clasps his hand and assures him of her presence. She reminds him that she promised she would be present when he returned from surgery. Seeing Christa above him, David's anxiety subsides; and he gratefully acknowledges her devotion to his care and responds,

"Thank you for being here, Christa. You are a welcome and beautiful sight to witness after my surgery. I now realize the importance of someone in my life that is concerned for my welfare. I hope you understand I am becoming very dependent upon your care and your presence for my wellbeing."

Christa acknowledges David's words and remarks,

"I know David. I am here, and I am concerned…very concerned for your welfare and full recovery."

David then inquiries about the duration and success of his surgery. He comments that his last conscious memory was being transported into surgery by a surgical orderly. Christa informs him that the entire surgical procedure took over six hours and was followed by an additional two hours of intensive surveillance and monitoring in postsurgical recovery. She was with David in postsurgical recovery while he was unconscious to confirm and monitor his physical condition. He was lightly sedated in postsurgical recovery for another 24 hours. She then accompanied David with the two orderlies who returned him to his Hospital Suite this morning.

Christa then radiates with gratitude and satisfaction as she declares joyfully,

"Oh David, the surgery was very successful as confirmed by your chief orthopedic surgeon, Dr. Feinstein. I am truly grateful for Dr. Feinstein leading the surgical team. He is among the finest orthopedic surgeons in America and probably throughout the world. Your injuries were very severe, and the intricate, surgical procedure was extremely demanding and delicate. However, Dr. Feinstein informed me after the surgery that your skeletal bones were all properly realigned, your tendons were successfully repaired and reconfirmed, and the peripheral nerves will regenerate with continued serial microsurgeries. These complex and complementary procedures are designed to restore complete muscular control, structural competence and full feeling to your right leg. The successful surgical restoration of such damage you sustained requires complex, delicate and demanding surgical procedures. Great medical skill is required to repair severely damaged, tissue, muscle and bones. Such surgical knowledge is only gained after decades of training and experience. Successfully executing such a demanding surgical performance is a surgical work of art, and Dr. Feinstein is the consummate artist for that class of surgeries."

David smiles at Christa as she concludes her postsurgical report and again gently squeezes David's hand. Christa ends her words by declaring to David with satisfaction,

"Please observe David, that my prayers in your behalf for the success of your first surgery were indeed answered. Both the Lord and I are concerned for your recovery and welfare."

David responds to his nurse with a simple,

"Thank you, Christa, for your prayers in my behalf. Your God hears your prayers."

Christa examines David and hints of marriage (8Oct Friday)

Later, during her Friday shift, Christa approaches David and asks him to roll on his left side so she can examine the surgical dressing applied to his right hip and leg during his surgery yesterday. She states she will replace the dressing if there has been exudation of serous fluid from the long suture line.

The dressing is slightly damp with serous fluid, so Christa gently removes the soiled dressing. She carefully examines the surgical area for evidence of continued drainage and potential latent infection. Observing that his sutures exhibit no evidence of infection, she then replaces the dressing. David is engrossed by her gentle, careful management of his severely injured hip and leg. David realizes that Christa displays care and tenderness for him that he has never experienced before, even from his mother. Such gentleness and love displayed by anyone in David's behalf is unfamiliar to him; but he now recognizes how greatly he appreciates and needs compassion and tenderness from a woman, especially from Christa.

As Christa completes her examination and replaces his surgical dressing, David realizes that she has observed, examined and treated all his body during his extended stay in the hospital.

Christa then disposes of the soiled bandage, thoroughly washes, dries and sanitizes her hands and returns to his bedside. David smiles at his compassionate and very attractive caregiver and asserts.

"Christa, you have examined me, robed and disrobed, changed my medical dressings, maintained my medical monitors, and even disposed of my rather unpleasant body eliminations."

Christa's causal and perfunctory response to David's declaration is the usual, obligatory response,

"That is part of a nurse's expected duties, David."

David then whimsically and provocatively declares with marked intent,

"It would be equitable if I had the opportunity to care for my beautiful nurse someday as tenderly and intimately as she has cared for me."

Christa is amused at David's provocative words and quickly responds with playful banter and obvious intent,

"My dear Professor that you could do only if you were to marry her!"

Christa winks at David as she utters her resolute and tempting declaration. David hastily responds to Christa's enticing words with a provocative question,

"That seems reasonable and very appealing…but would my nurse ever consider marriage with her patient? Can a devout Christian woman, a Mormon, marry a secular, Jewish man?"

Christa answers immediately and with complete and emphatic assurance.

"Certainly! If they love one another and they both accept and love the Savior, Jesus Christ."

David silently ponders Christa's emphatic declaration as she resumes her nursing surveillance for David.

As Christa performs her duties, she now realizes that this simple and brief verbal interchange has broached her marriage with David, and established Christa's essential prerequisite for their mutual belief in Christ. She has assertively sown the seed for their future marriage in David's fertile mind and unfulfilled heart.

Later in this day's shift, as Christa continues her nursing duties for David, she deliberates on their earlier verbal exchange. Would David consider marriage with her, a Nurse...a Christian...a Mormon? Christa, recalling Dr. Feinstein comments during David's first surgical procedure, silently ponders this remote possibility with wistful delight. Reflectively, perhaps with continuing interaction and nurturing, enhanced by familiarity and companionship between her and David, their close association may develop into a binding, sacred and intimate relationship, culminating in marriage and David's membership in the Church. Still, Christa recognizes this outcome is very improbable. She has known David for only one week and she is simply an assigned nurse, temporarily caring for a patient...a very famous, non-Christian, Jewish scientist. Is she simply infatuated with her newly assigned patient?

Christa's Sunday events (10 Oct Sunday)

After attending church at the Cambridge Ward, Christa makes her regular Sunday evening long distance telephone call to her mother at Hospice Cancer Center associated with the Huntsman Cancer Center in Salt Lake City, Utah.

Closing her call with Christa opens her Journal and reviews her earlier entry for last week. With her mother dependent upon Christa's financial support and the demands of a full schedule of nursing duties at Massachusetts General Hospital, Christa has had an unfilled social life. Her principal social interactions outside her Hospital environment are confined to active attendance at the Cambridge Ward and occasional Temple Sessions at the Boston Temple with Christa's Relief Society Sisters in the evenings and an occasional Saturday Session. Now, however, the advent of her full-time assignment and concentrated care and attention for David Steinmann has aroused her interest in men, particularly David. Christa ponders pensively upon marriage and a future family of her own. As Christa contemplates these issues, she lifts the pen from her small kitchen table and records the following words in her journal.

Christa's weekly Personal Journal for 4-10 Oct 2010

David underwent his first surgical procedure this week, and it was very successful. I am very grateful for the skills of the surgical staff, particularly Dr. Feinstein, and David's excellent medical progress. My sustained prayers for his full recovery are being answered in his behalf. I am truly grateful and blessed, Heavenly Father.

I believe David likes me, and I know that I care for him. Dr. Feinstein is very perceptive regarding my increasing interest and growing affection for David.

David is a handsome, young Jewish Scientist, and I do enjoy being his nurse. This assignment is an honor for me and a great responsibility to contribute to restoring David's injured leg so that he can walk again and pursue his important scientific research.

But what will be the significance and consequences of his request for my exclusive nursing service and our expanding, intimate association with each other? I wonder. We have even broached marriage in our conversations. What does the future hold for David and me? What does the Lord have planned for my future and David's future? Guide me, Heavenly Father that I may know what my interest and relationship with David can be and should be. Do David and I have a future together? Can I bring him to the Savior and then to me as his wife and eternal companion? Please, Heavenly Father, guide me through the promptings of the Holy Ghost.

CHAPTER 4 (11 Oct-17Oct)

David's acute sensatory perception (11 Oct Monday)

This Monday morning, as Christa begins her daily nursing shift and closely monitors David's medical probes to assess and document his recovery from his first surgery, David offers a surprising assessment of Christa as she performs her usual duties near his bedside.

"Christa, while you are examining and assessing my recovery from surgery last week, let me perform my own inspection and assessment of my beautiful and competent personal nurse as she commences this day's activities in my behalf."

Curious and aroused at David's invitation, Christa smiles ardently and approaches closely to his bedside, places her hands on her hips and asserts provocatively,

"I am prepared for your personal analysis and diagnosis; Dr. Steinmann...please...do begin."

David gives his comprehensive appraisal of Christa's morning.

"Well, Miss Olsen, for breakfast, you ate a grapefruit and an oat-based cereal with nonfat milk. You do not drink coffee, but you did prepare and consume a hot chocolate drink...probably a Carnation Brand. All these activities occurred while you were still in your Church garments, which I learned from the Internet that devout Mormons wear. Then you showered with Dove Soap, donned your attractive frame with fresh Church garments, and lightly applied some perfume. I do not know the brand, but it has a lilac flower base, probably a favorite fragrance. Your underarm antiperspirant is Secret for Women. You then brushed your teeth with Colgate Total Brand toothpaste and donned your white blouse and your attractive, form-flattering skirt that you had laundered in Tide. You then gently brushed your long, golden blond hair and fashioned that silky bundle into that flowing ponytail that I intently monitor as you elegantly move through this Hospital Suite. Slipping into your nursing shoes, posting the breakfast dishes in the dishwasher and placing a Finish Powerball in the detergent receptacle, you secured your purse, locked your door, and quickly left your apartment passing a kitchen where the tenant had burned the breakfast toast. You briskly walked to the bus stop where diesel exhaust from the bus tainted the atmosphere while you were boarding the bus. You completed the trip to the Hospital on MBTA, where you sat by some other medical personnel who probably work in the Massachusetts General Hospital morgue...a slightly detectable scent of formaldehyde...sorry about that. You then stopped at the Hospital Pharmacy before coming to my Hospital Suite and obtained my specially formulated sedative formula that accommodates to my history of epilepsy. It is that low dosage codeine-based prescription for me that I dislike since codeine derivatives give me headaches."

Christa is astounded and dubious over David's statements, but she realizes he is minutely precise and accurate in each declaration. She challenges him and replies that he simply made a fanciful, but surprisingly correct assessment of her activities before arriving at David's Hospital Suite. However, David declares in response to Christa's claim,

"My highly developed sensatory ability is really not inexplicable or truly unique. Animals, especially canines have a highly developed sense of smell and, after all, men are animals. There is recent medical research evidence that canines can even detect the presence of certain cancers as they occur and metastasize in humans. Furthermore, skilled professional coffee and wine tasters can discern and identify hundreds of different scents and aromas associated with their aromatic products. These tasters can tell the origin, and even the location and harvest season for their products."

Christa acknowledges her patient's response, but then authoritatively and sternly retorts,

"Although my olfactory perception is not nearly as developed as is yours, I have no difficultly recognizing when your urine bag requires a change. Furthermore, David, men are not animals! You are a Spiritual Son of God who is our Father in Heaven. He loves you and He loves all of His Spiritual Children."

With Christa's stern, rebuking words, David is becoming aware of how deeply and personally, Christa embraces her faith in and reverence for her Christian God. David realizes he must exercise judgment and caution in conversing about issues that address religion and deity with Christa.

Christa will learn that David has a very keen, highly developed sense of smell that David slowly, but effectively, confirms to Christa. Indeed, Christa later discovers that David held a research contract with DHS, the Department of Homeland Security, to investigate and develop a quantitative model to operate a specially designed integrated circuit microchip for the olfactory perception of certain chemical compounds, especially explosives used by terrorists. The very sensitive microchip detector can respond to a single molecule of a targeted explosive compound.

David frequently demonstrates his highly developed human sensatory faculty to Christa, sometimes to her embarrassment. He can tell what she has eaten, where she has been, and the aromatic environment surrounding her presence. Furthermore, David can accurately assess the biological state of Christa's body, characterizing her 'unique fragrance' as a delicate, balanced bouquet of feminine aroma and female bouquet. He artfully depicts Christa as a masterful, fragrant blend of femininity and desirability.

Embarrassing to Christa, David can also tell when she and other female visitors to his hospital suite are ovulating and menstruating. He playfully enjoys enlightening Christa about her calendar schedule of female biological functions. She is doubtful and even embarrassed at first, but soon learns the accuracy of his assessments and reluctantly acknowledges and eventually supports his unusual sensatory gifts.

David and Christa's intensifying interactions together

Thus, continues a sustained repartee and banter of verbal and social exchanges between David and Christa as she provides close, personal medical support and monitoring of David's medical treatment at Massachusetts General Hospital. She is in David's private Hospital Suite each weekday ensuring that he receives essential nursing care. Christa also supports his research efforts. She helps him to use his computer and locate notes and materials to pursue his research. She is continually tasked with accessing his computer printouts from the wireless laser printer in his Hospital Suite and retrieving the papers, mechanical pencils and erasers that fall from his bed and table as he writes, deliberates, corrects, and rewrites his equations, calculations and notes. Christa is directly experiencing the iterative, protracted processes that dedicated scientists, like David, perform as they seek to discover new concepts in science and explore their consequences.

David often shares his work with Christa in scientific discussions and inquiries. David is impressed with Christa's mental acumen and superior, innate intelligence. Her rapid and astute assessment of David's scientific inquiries challenges him to explain difficult and obscure concepts until they are understood by Christa and fully transparent to David. David realizes that if he can demonstrate the logic and correctness of his models and theories to Christa, a non-scientist, then others will readily comprehend his work and be persuaded to accept his labors.

Christa's support of David's appearance (12-15 Oct Tuesday-Friday)

In addition to supporting his research and insuring his medical care and participation in physical therapy, Christa is also sensitive and assertive about David's appearance and surroundings. She ensures he is attractively groomed and presentable for the many visitors he entertains from Harvard and other universities who confer with him regarding his research. Christa is constantly alert to potential exposure of David to viruses and pathogens carried by visitors. She rigorously mandates and enforces sanitation and hand washing by visitors before they physically contact or exchange items with David for his use. Occasionally, she intervenes when the visitors overextend their stay, and Christa recognizes that David is becoming overstressed and requires restraint and rest. Although David gently complains over Christa's protective oversight, David recognizes her concern and care for him and appreciates the time and careful attention he garners from this beautiful, committed caretaker who now has a dominant presence and crucial role in his life.

Christa insures David's bed linens are changed at least daily, and she brings fresh flowers to his Hospital Suite each week usually on Tuesday morning. She provides all his nursing needs...ensures a pathogen free environment...acquires and administers his medicines...monitors and records his biological data and maintains his medical records. She also provides the essential exchange of information and assessment with the Hospital physicians and staff who frequently visit his Hospital Suite for examination, observation, consultation and assessment of their famous Harvard patient.

Most important, Christa is responsible to prepare, support and sustain David for the extensive tests, CT scans, and serial surgical procedures required to assess and repair his severely damaged left hip, thigh and leg. Although Christa is kind and considerate with David and respects his scientific genius, she authoritatively requires that he complete his regular, intensive physical therapy schedule even when he complains that the therapy session interrupts his research. Christa must frequently remind David that she is responsible for his physical welfare and medical recovery and she will not compromise that obligation.

Christa's duties in David's behalf are demanding and intense, but very satisfying for her. She realizes that she has assumed the dominant role in David's life for insuring his physical recovery and wellbeing. Gradually and imperceptibly, they are bonding as a couple as Christa overtly realizes and intends to transpire. Christa is establishing the supporting role she would wield if she and David were married.

Christa's loses her apartment (15 Oct Friday)

On this Friday morning as Christa begins her daily nursing duties in David's Hospital Suite, Christa is obviously troubled and irritable with the other nursing staff and occasionally even with David. David immediately recognizes her latent distress and gently inquiries about her anxiety. However, Christa is reticent and reluctant to respond and to engage David in her usual playful banter. Finally, David must act and declares emphatically,

"Christa, you are not the witty, engaging nurse today with whom I delight and require for my nursing care. I have spent a dull, lonesome night in this hospital cell and I need your attractive and stimulating presence to enliven me and brighten my day. Today is Friday, and I will not see you again for three days. Something is distressing you, and I must know what it is. Please tell me; what is the problem?"

Christa responds curtly, stating,

"David, it is a personal matter and I will not encumber you with the issue. Besides, there is nothing that you can do other than commiserate with my frustration and plight."

Her serious declaration now greatly concerns him, so David continues to press Christa for disclosure of the troublesome issue. Finally, after considerable cajoling and gentle prompting by David, Christa responds reluctantly,

"David, I have lost my entire year's lease payment due to the mortgage default of the previous apartment owner where I live. The new apartment owner's bank will not honor my previous rental agreement and lease payment. Their bank demands another full one-year lease payment if I want to retain my apartment. The bank contends that the mortgage default issue is in litigation for recovery of certain financial obligations of the previous owner. In the interim, I must tender another year's lease payment to retain my apartment. I do not have the funds to cover that unexpected expense. My father is deceased, and mother is a terminal cancer patient in a Medical Hospice Facility in Utah. I must augment her Social Security and TRICARE military dependent benefits to cover her medical and living expenses. David, I am desperate. This unexpected, additional financial obligation poses overwhelming issues for me, and I do not know what to do."

Then pausing, Christa soberly states,

"David, I am sincerely sorry for burdening you with my problem. You must concentrate on recovering from your accident and restoring your right leg."

David requests Christa occupy his apartment (15 Oct Friday)

David, recognizing and sensitive to Christa's plight, responds quickly with the obvious solution as he advises,

"Christa, the resolution of your problem is simple…abandon your apartment and move into my apartment. My apartment is empty; I have an irrevocable two-year lease; and the apartment requires maintenance and upkeep. I must pay the rent for the apartment whether it is empty or occupied. The apartment is in disarray, as you must know from your previous visits. You have visited the apartment to secure my research materials, and you know the apartment needs attention and, importantly, occupancy and care. In addition, I will need you to continue to access my materials there, so I can continue my research."

David then with a satisfied grin retorts quickly as an astute, scientific advisor to his beautiful, but distressed nurse,

"Ergo, Miss Olsen, the solution to this simple problem in contentious, domestic contracts follows immediately from your declared facts. Your occupancy of my apartment is the practical and obvious solution. Logically, your occupancy is the necessary and sufficient solution for me and especially for you."

Christa frowns and responds saying simply,

"No David, I cannot occupy your apartment. Your logical conclusion ignores the moral imperative."

David again importunes her now, strongly imploring,

"Christa, please occupy my apartment. You need a place to live and I need you to maintain the apartment and bring me my mail and access my scientific materials."

Christa remains averse to occupying David's apartment and finally declares to his persistent requests,

"David, a religious, moral woman does not occupy the apartment of a man who is not her husband. What will your university contacts and neighbors believe and say?"

David, without careful consideration, immediately responds defensively,

"Well, Rebecca shared the apartment with me so simply tell those moral antagonists that you are my other sister, and you will now be occupying the apartment."

Christa immediately retorts indignantly with the disclaimer,

"A devout Christian, a Mormon woman, does not lie! And as a Swedish, fair-skinned, blue eyed blonde; I am an unlikely sibling for you, Doctor Steinmann."

David pauses, reconsiders his previous, impulsive words, and then thoughtfully responds reflectively,

"Yes, Christa, you are correct. You are far too lovely to be my sister, and I am very glad you are not my sibling. My interest in you greatly exceeds sibling sentiments and desires."

Throughout the Friday shift at the Hospital, David and Christa continue to offer and oppose reasons regarding Christa's occupancy of

David's apartment. Christa expresses concern that David's mother, Sariah, may come to the apartment to secure Rebecca's belongings. How would Christa counter his mother's condemnation and rebuke if Christa were now living there. Would Sariah denounce Christa's moral character? The accusations might escalate to the level that Sariah would seek Christa's discharge from the Hospital and removal from David's nursing care. Moreover, moral charges against David would harm his reputation and career at Harvard University and even his research endowment.

This exchange of supposition, assessment, rejection, and counter offers continue throughout the day between David and Christa regarding her occupancy of David's apartment. Finally, after Christa and David have exhausted verbal rhetoric addressing this issue and as Christa is concluding her nursing shift for the week with David, he gently pleads,

"Christa, I will telephone my apartment manager and ask for his permission for you to occupy the apartment during my extended absence. I will ask the manager to prepare a new, sublet lease for your occupancy. You would then simply be occupying my previous apartment as the new tenant. If the manager agrees, will you then at least consider occupying the apartment?"

Christa, weary from their protracted exchange, says she will consider David's proposition as she departs Massachusetts General Hospital for her own apartment for the weekend.

Christa's Sunday's Journal (17 Oct)

During the weekend, Christa struggles with alternative practical and respectable resolutions for this situation. She has visited David's apartment many times to secure his research materials and mail. She is familiar with the apartment tenants and has befriended David's neighbors who often inquire about David's welfare and the restoration of his severely injured leg. Christa has graciously addressed their concerns and assured them of David's continuing medical progress and improvement.

Christa makes her weekly telephone call on Sunday to her mother in Utah and confers with her regarding her judgment about David's offer for Christa to occupy his apartment. Christa has previously discussed with her mother her assignment as David Steinmann's permanent nurse. David will be a patient and confined to the hospital for several months while extensive surgeries and physical therapy are performed to repair his severely injured leg. His apartment will be vacant during that time. Christa and her mother mutually deplore the unjust treatment by the new owner of Christa's present apartment. However, Christa's mother is also concerned over Christa's occupancy of David's apartment and the possible impact upon Christa's reputation. Finally, as they conclude their weekly telephone conversation, her mother advises Christa to pray regarding the proper choice and allow the Holy Ghost to guide her to the proper decision.

Later that Sunday evening, Christa solemnly ponders her options for residency and tacitly appeals to the Holy Ghost for guidance. She then records the following entry in her Journal before she kneels for her evening prayers and retires.

Christa's weekly Personal Journal for 11-17 Oct 2010

I have lost my apartment lease. What can I do? What should I do? I do not have funds to repay another yearly lease where I live. How will this unwarranted act by the new apartment owner affect my duties at the Hospital and my financial responsibilities for assisting in my mother's medical care? In addition, will this unexpected financial burden affect my ability to support my mother and compromise my duties as David's nurse? Should I secure another position at the Hospital or elsewhere in medical practice that would increase my income? Would my bank grant me a loan to cover the additional lease with the hope that the legal issues confronting the present and previous apartment owners would be quickly and fairly resolved in my favor?

David has offered me his unoccupied apartment. However, can I accept David's offer? Would it be morally and ethically wrong and affect my reputation and David's reputation and career if I occupy his apartment?

Father in Heaven, I need Thy wisdom and approval if I should do as David requests. I have plans and aspirations for David and me and our future together, and I do not want to quit as his nurse and find another assignment at the Hospital or employment elsewhere. However, will my residence in his apartment threaten those plans? I am prepared to follow Thy will, Heavenly Father, please guide me as to my actions.

CHAPTER 5 (18 Oct-24 Oct)

Christa occupies David's apartment (18 Oct Monday)

With sustained, but gentle persuasion and Christa's inability to raise another year of lease payment, David prevails in his repeated requests that Christa occupy his empty apartment. David has called his Apartment Manager who has met Christa several times at David's apartment and will welcome her as a tenant. The Manager declares his approval and the need to secure the apartment, and he seeks her immediate occupancy. Christa's tenancy will be formally recognized as a sublease of David's apartment. Christa also realizes that David's offer is evidently the Lord's answer to her prayers. So, Christa agrees; she will move into David's unoccupied apartment during the week. She will occupy the apartment while he is hospitalized for his indeterminate stay. Then upon David's release from the Hospital, Christa will make other arrangements for housing...unless, of course, she becomes David's wife.

After her Hospital shift on Monday afternoon, Christa travels to David's apartment and informs the manager and other building tenants that she will occupy the apartment. She signs the apartment manager's Tenancy Agreement and receives additional keys and information to access other facilities at the apartment complex. Much to her relief as David reported, the manager and other tenants graciously welcome her and will assist her with moving her limited furniture into the apartment. The apartment tenants are pleased that they will now receive frequent reports on David's recovery at the Hospital. Furthermore, the manager and neighbors are relieved that the empty apartment will now be occupied and secured and be better maintained. In addition, to have a skilled, registered nurse in the apartment complex will provide many of the senior residents with security and peace of mind in the event of a medical emergency.

During the week in the evenings, Christa progressively moves her limited furniture and other belongings into David's apartment. She packs, labels, and stores Rebecca's remaining clothing and personal items in the secure storage area at the apartment complex. Christa will occupy Rebecca's former bedroom.

Christa is heavily tasked in the evenings after her shift at the Hospital with cleaning, rearranging furniture, doing laundry, storing her clothes and personal items, and general housekeeping in David's apartment. With David's permission and suggestions, she skillfully reorganizes and restores the vacant and neglected apartment into a comfortable and well-organized residence. She imposes her feminine touch upon the appearance and layout of the apartment. Subtly, but consciously, Christa realizes that she is reorganizing David's apartment into a residence as she would do if she and David were married.

Christa assumes David's personal affairs (18 Oct Tuesday)

While cleaning, rearranging and living in the apartment, Christa recognizes that although David is an acknowledged scientific genius, he is a poor housekeeper and inattentive manager of his apartment and affairs, particularly his personal and financial affairs. He has little interest in domestic and financial organization, which he assigned to Rebecca. Also, as David's continuing financial obligations become due and arrive in the mail at the apartment, unpaid, Christa expresses concern to David over proper disposition of these past due notices. In response to Christa's concerns, David implores Christa to handle his financial affairs while he is hospitalized and incapacitated. David confesses that Rebecca serviced all the financial affairs for both of them. He has neither the time nor interest to deal with these domestic issues. He must concentrate on his research that has been unduly delayed because of the automobile accident. Continued support of his Endowed Chair position at Harvard is dependent upon his successful development and acceptance of M-Theory by the scientific community. David repeatedly entreats Christa to perform those necessary but tedious responsibilities for him as Rebecca had done.

Christa is reluctant to assume those financial duties that usually belong to a married couple; but David implores Christa to assume these tasks at least while he is hospitalized. As David continues to receive overdue payment notices and persistent telephone calls at the apartment, Christa finally relents and assumes these responsibilities. David petitions and gives Christa joint signatory authority for his checking account, so she can manage his financial affairs. She recognizes that David is unable, or at least unwilling, to assume these duties while he is hospitalized; and creditors may cause him unnecessary stress. Gradually, Christa fully assumes and executes the role and responsibilities that Rebecca provided David by handling all his personal and financial affairs.

Provocatively, Christa now realizes that in many ways she is David's surrogate wife for his personal and financial affairs. Although she verbally expresses reluctance and concern to David for her escalating and dominant role in managing his personal affairs, she is secretly pleased since this vicarious role delights her. Christa recognizes that David's dependence upon her is imperceptivity forging their union. Christa is also discovering and experiencing what her spousal duties and lifestyle with David ben-Steinmann, the renowned scientist, would be if they were married.

As she performs these mounting duties and responsibilities for David, Christa carefully and frugally handles both of their financial affairs. She meticulously provides a full accounting of all income sources and expenses to David, even though he shows little interest in his financial state and declares his full trust and confidence in her decisions and actions. However, Christa insists that she maintain her personal finances and checking account independent of David's bank account, since his income far exceeds her income. She insists that she must provide for her own personal needs. David recognizes and acknowledges her careful and frugal care for his affairs. In fact, to David's delight, as Christa handles and accounts to him for his finances, he realizes his economic condition is now actuarially sound and has markedly improved with Christa's management. David also recognizes his growing reliance upon Christa in his life and the important role she has now assumed for all his fiscal and domestic affairs as well as medical management. David's dependency on and respect for this Mormon, female "enchantress", is increasingly evident to him.

Christa obtains a BlackBerry

David finds that his relationship development and time with Christa is limited if he can only communicate with her when she is physically present in his Hospital Suite. David has a Model 9780 BlackBerry with 512-Megabyte memory and an eight-Gigabyte media card with MPEG4 Video including a 480 by 360-pixel color display and MP3 Audio. The BlackBerry is fully outfitted with mobile telephone service, email, texting, a camera, and a GPS. If David can persuade Christa to acquire an identical BlackBerry and add it to his Calling Plan, then he can access her by voice, sight, and email wherever she may be. He realizes that she may object to such continuous interaction and surveillance during her off-duty time, but Christa is now the dominant and essential person in his life. He needs her service and support to continue his research and more importantly, he wants her full presence in his life. His growing dependence and reliance upon this lovely blond, blue-eyed, young Mormon woman is increasingly evident to David.

During the noon hour, when Christa brings David's lunch to his bedside and they eat together, he inquires of her,

"Christa, I know you have an old AT&T cell phone, but it has limited capacity and few options for communication modes and use. Would you consider adding a BlackBerry to my calling plan for our mutual use? We could then communicate with each other at will wherever you may be. The BlackBerry offers mobile telephone, email, an on-line camera, texting, a GPS, Google, and many other features including the internet, Gmail, and APS that are very useful. Often, I need you when you are not here at the Hospital Suite. Also, when you are at other locations in the Hospital I could call you if I need you. There are many scientific references and other research materials in the apartment that I require. I often realize the need for those research materials when you have departed for the day. Frankly, Christa, I miss your presence when you are not here at the hospital, and mutual BlackBerry phones would provide us audio and visual access with one another."

David continues in his subtle and disguised petition to convince Christa to acquire a BlackBerry phone, so he can access her and have quasi-continuous communication and interaction with her. He fortifies his petition saying,

"Furthermore, when I have questions regarding Scriptural passages you could immediately respond and provide me with your insight and beliefs. I realize that you are a serious student of the Scriptures, especially those Mormon Scriptures that I may need your assistance in understanding. Christa, I especially need and would welcome your personal testimony on a continuous basis. If you want me to embrace your Savior, I need your constant input and strong testimony to guide and encourage me."

"You could use the camera on the BlackBerry, and I can direct you to the correct location for the research materials I need from the apartment. In addition, I would like to see what you have done with the apartment. I understand you have reorganized the apartment and made the rooms more habitable and inviting. I would particularly like to observe what you have done to my disorganized and chaotic bedroom and the cluttered gathering of books in my bookcase."

"Frankly, Christa, I miss not seeing you in the evening. The truth is I need more of your presence in my life. This Hospital Suite is very uninviting and lonely when you are absent. What do you say to my proposition, Ms. Olsen?"

There is a long pause with no response by Christa. Finally, she counters,

"David, let me ponder your offer. I recognize your need to identify and access your research materials and other items in your apartment. However, I am a private person and have lived alone for many years. To have you constantly accessing me is a new dimension in my life that I must consider. However, you are correct, there are many times when I am also lonesome in the apartment. To have your presence, but with controlled access, would be challenging and satisfying. Perhaps we can continue our religious discussions more frequently and to greater depth. We would not have the constraints that exist in your Hospital Suite like extended Gospel discussions and even praying together. I do want you to learn more about the Gospel, the Gospel of Jesus Christ, that is essential in my life. However, David, we must establish ground rules. The BlackBerrys are off limits when I am dressing or undressing or in the bathroom or even need to be alone."

David responds quickly and with fervor.

"Agreed, and I acknowledge and will honor all your conditions and restrictions. You will have full control over your BlackBerry and my access to you. However, I exercise no privacy or modesty from you, Nurse Olsen. Of course, all your personal tasks in my behalf are 'in the line of duty'."

Christa counters briskly and professionally,

"David, you are my patient, and I am responsible for your health and recovery. Remember, you requested my exclusive commitment for your care, and my priority is to make you ambulatory again. I am tired of parading my handsome, famous young patient throughout the corridors of Massachusetts General Hospital in a wheelchair. I want you to walk the hospital corridors with me, perhaps even holding my hand. Finally, I want you to walk, unaided from Massachusetts General Hospital, when you are discharged."

Satisfied and delighted with Christa's acceptance of his proposition, David closes the discussion by eagerly acquiescing.

"I gratefully accept all your conditions and restrains. I would be an isolated, lonely man without you and your support. And to walk with you through the Hospital corridors, even with my cane, will make me an envied male patient at Massachusetts General Hospital."

David then tells Christa the location of the electronic shop near Harvard where he acquired and services his BlackBerry. He will call and inform them what BlackBerry Model, calling plan, and accessories that she will need for their mutual electronic communications to function properly. Her BlackBerry will be a Model 9780 identical to his.

As the Friday nursing shift closes and Christa prepares to return to David's apartment, David reminds her to secure the BlackBerry over the weekend that he has selected and charged to his account. As she prepares to leave David's Hospital Suite, Christa perceptively responds.

"Dr. Steinmann, I believe that these BlackBerrys may introduce a new dimension to our relationship, although we will remain constrained in our present four-dimensional, spacetime universe."

Christa continues her concerns with some anxiety,

"I hope you will not scrutinize and critique my housekeeping as carefully as you do your scientific and mathematical endeavors. Moreover, please condone my idiosyncrasies and habits in housekeeping. I have lived alone since graduating in nursing from Brigham Young University. My housekeeping style and agenda are my own and firmly established."

David counters Christa words saying,

"Christa, if your homemaking skills at our apartment and your proficiency with our budgetary affairs are reflected by your competence and skills here at the Hospital then I have acquired a precious pearl. I believe you refer to such an acquisition, using the language of Mormonism, as a 'Pearl of Great Price'. Christa, understand that I want your presence in my life, stronger and longer. Do you understand the depth of my dependency and affection for you?"

Christa's pale blue eyes now focus intently upon David as she replies ardently,

"My growing need for you in my life is also deep and compelling, David. Yes, I will acquire a BlackBerry and bring it to you on Monday for inspection and personal instruction."

With these closing words, Christa departs from David's Hospital Suite for the weekend.

David's assessment of Christa (Oct Saturday)

During the following weekend while Christa is absent from David's Hospital Suite, he ponders over his assessment of Christa and the dominant impact and role she has now assumed in his life. David cerebrally evaluates his factual data and careful evaluation with the following serial, precise, quantifiable assessment of Christa Olsen, RN as Massachusetts General Hospital.

"Christa is a beautiful, statuesque, 28-year-old female registered nurse at Massachusetts General Hospital and, most important, unmarried. Christa' physical features are stunning and provocative. Christa's waist is very small; I estimate the abdominal circumference is about 50 centimeters. I must modestly retreat from speculating on her breast measurement, but it is obviously ample and pleasing. I am continually preoccupied with Christa's demeanor and unique feminine form as she expertly performs her nursing duties in my Hospital Suite."

"Christa has long slender fingers. But her fingers and hands exhibit strength and dexterity as she works. She skillfully uses her hands to perform the numerous tasks, both necessary and unpleasant, that are associated with expertly caring for me. She must routinely examine and change my surgical dressings, empty my bedpan, maintain and confirm the patency of my medical lines, monitor my medical information and administer my drugs. She accomplishes all of these tasks, even the distasteful ones, with efficiency, gentleness and without complaint."

"She has golden blond hair that is coiffured into a prominent ponytail that hangs gracefully as a dorsal appendage from her head. This bouncing bundle of blond silk flirtatiously sweeps from side to side across her slender swanlike neck as she moves and performs her nursing duties. I am fascinated with her refined social bearing, her graceful movement, and especially her consummate, physical beauty."

"Aside from her corporeal beauty, Christa is very intelligent and joyfully playful. She can banter with me and rapidly counter, my provocative words as well as I can deliver. Her flawless complexion, high cheekbones and large, pale blue eyes crown her elegant, angelic face and mirror her emotional moods and attitude. Her sculptured nose has a slight upturn that endows her with an aura of youth and innocence. All these physical features complement and consummate her statuesque profile and stunning pulchritude."

"My conclusive assessment is that Christa is a very bright, slim, statuesque female, who has perfect complexion, large pale blue eyes, blond hair set into a pirouetting pony tail, a superior mind, a keen sense of humor, and is willing to engage and banter with her patients...at least with me to my immense delight and continuous enjoyment."

"Christa is determinedly optimistic with and for me. She has an innate sense for assessing my mood and can elevate my spirits and provide my physical, emotional and mental needs. Her presence in my hospital suite assuages the transient pain I experience from my severe injuries. Christa has convinced me that I will fully recover from my injuries to my leg, and I will be able to walk again. Because of her nursing skills, tender care, and confidence and optimism, she is the critical entity to restoring my health."

"I can discuss and even explore profound scientific issues and obscure scientific concepts that Christa rapidly comprehends and then challenges me with remarkable insight. I find that in presenting my scientific suppositions to her, she initiates the necessity for me to clearly explain and make credible and explicable my concepts and tentative theories in science."

"On the personal side, I can confide in her and express my own emotions and needs. This is a very unusual gift Christa has that I have not experienced with my own family and especially my mother. I am completely comfortable and secure in disclosing my personal feelings and concerns to her that she will honor and respect. In summation, I wish that Christa were present in my life...permanently; but only if I can recover my full mobility and strength. I cannot and will not burden Christa with a handicapped companion."

Christa's assessment of David (23 Oct Saturday)

Coincidentally, during the same weekend while she is absent from David's Hospital Suite, Christa ponders her own assessment of David. Christa contemplates her evaluation for entry into her Journal that she will make on Sunday evening. David and she will share BlackBerry cell phones and this experience will greatly increase our interaction with each other. As Christa ponders her appraisal of David Steinmann, she tacitly expresses the following images in her mind.

"I am greatly impressed with David and his immense professional status and achievements at such an early age. At 30-years of age, he is an internationally recognized, scientific genius who fascinates me with his wit as well as his wisdom and compassion. With his dark complexion and black, thick, curly hair, he is a handsome and charming young man, as well as brilliant physicist. He is possibly destined for a Nobel Prize and a significant place in the field of astrophysics and cosmology. I am flattered that he has sought me for his solitary medical care. I greatly enjoy serving him and engaging in pleasant verbal banter with him. Although demurely, disparaging his conspicuous attention, I delight at his rapt observation of me and his ubiquitous compliments. It is my responsibility to ensure the successful and rapid recovery and return of this young Jewish genius to the esoteric world of physics and cosmology... where he may alter the entire field of science."

Then with an inner gleam, Christa mentally muses further, "...and where he may transform and bless my life as his eternal companion if I can bring him to the Gospel as a devout Disciple of Christ."

Christa's weekend activities (24 Oct Sunday)

After Church on Sunday, Christa is now fully situated in and comfortably acclimated to David's apartment. She realizes that her accommodations in this new setting are far superior to her former apartment, although David's apartment was initially very disorganized and required thorough cleaning and remedial housekeeping. Christa occupies Rebecca's room and Christa has rearranged the room to meet her needs. Christa has carefully packed and stored Rebecca's clothes and personal belongings since Christa realized that David's mother may want these items returned in the future. Included In the personal belongings is Rebecca's Diary and a checking account in David and Rebecca's name. At David's request, Christa cancelled Rebecca's checkbook that was a joint account with David. Christa has also saved Rebecca's other personal effects for delivery to David's Mother.

David's room required considerable attention to organize his clothes, his books, and other research materials so that Christa could identify the research resources and bring them to David as he requests. She will pan all the rooms in David's apartment with the BlackBerry camera and seek David's input as she reorganizes the apartment. David has been unconcerned about the reorganization and has asked her to change the apartment as she wished. David's principal interest in the contents and arrangement of the apartment were his research materials including his computer files, scientific books and journals, and somewhat dispersed notes and calculations. These materials Christa has collected, carefully sorted, and filed for David to identify and for her to secure, as he needs them.

On Sunday, Christa makes her usual weekly call to her mother in Utah. Christa cautiously and carefully informs her mother that she is now living in her patient's apartment while he is in the Hospital. After protracted and persuasive explanations regarding her occupancy in the apartment of her patient who is isolated at Massachusetts General Hospital, her mother approves of Christa's action. Christa discusses at length, her assessment and growing appreciation and affection for David. After considerable discussion, Christa concedes to her mother that she has deep feelings for David. He is brilliant, kind, and considerate. Christa informs her mother that she loves David and wants to make David her husband.

However, Christa's mother is concerned that since David is Jewish, it will be very difficult to convert him to the Gospel of Jesus Christ. Christa must marry a worthy man who is a member of the Church. Christa's mother pleads that Christa marry David only if he is baptized, confirmed, receives the priesthood, and their civil marriage is later sealed in the Temple. Christa's mother strongly reaffirms her petition by declaring solemnly,

"You are the only child that your deceased Father and I have; and I must have you, your husband and your children sealed to us as an Eternal Family. Christa, you know that my remaining time in mortality will be short; and I must have your assurance that you will honor this binding expectation from both your father and me. We must be joined together as an eternal family when you marry. Christa, will you promise to fulfill our request?"

Christa answers,

"Yes mother, I will only marry a man who is or will become a member of the Church. I also want to be with you and my father in the Celestial Kingdom."

As Christa closes her weekly telephone conservation with her mother, she ponders her commitments to her mother to marry David only if he becomes a member of the Church. Can she achieve these difficult demands and goals? She loves David deeply and wants to marry him, but can she secure his membership in the Church and then have their marriage sealed in the Temple? When David is released from the Hospital, their future interactions with one another will be limited or may even end. David's research may dominate his time and interests, and she may lose him. Is his maturing love and need for her sufficiently developing and his experiences with his NDE, and the influence of the Holy Ghost adequate to initiate baptism and achieve all the expectations she holds for David and his acceptance of the Gospel to satisfy her promises to her mother?

Before retiring the Sunday evening, Christa reviews her previous journal entries since becoming David's nurse. She recognizes the dominant role that David now exerts in her thoughts and her heart. Removing the journal from the nightstand, she thoughtfully ponders the words for entry in her Journal. She then she pens the following entry:

Christa's weekly Personal Journal for 18-24 Oct 2010

In an unusual answer to my prayers, I am now living in David's apartment. David has cleared my residence there with his apartment manager. Everyone at the apartment complex supports my temporary residency while David is hospitalized. The apartment tenants appreciate my medical care for David and maintenance of his apartment in his absence.

David's apartment is large and comfortable, but it was very disorganized, especially David's room. I have tried to correct the situation and exercise my homemaking skills. In addition, I will acquire a BlackBerry that David wants for our sustained communication with each other. With our mutual BlackBerrys, he will now have nearly constant access to me here in the apartment. With the presence and use of the BlackBerry in our mutual lives, perhaps our relationship will develop to the point that David will marry me.

I am delighted with my new residence and believe that living here in David's apartment is a prelude to becoming David's wife and companion. Father in Heaven, I do feel Thy approval and support for my occupancy, and that my plans and aspirations for David and me are pleasing to Thee. I hope that this arrangement will result in our celestial marriage if it is Thy will. Please bless me so I can bring David to the Gospel of thy Son, Jesus Christ. With Thy approval, and the support of the Holy Ghost, I know I can achieve that blessing.

I have promised my mother to marry a man who is or will become a member of the Church. Therefore, I must nurture and secure David's baptism and his receipt of the Priesthood and our sealing in the Temple. Father, I believe I can accomplish that goal if that is Thy Will. Please bless me that my desire and aspiration for David are also Thy Will.

CHAPTER 6 (25 Oct-31 Oct)

Christa and David will share BlackBerrys (25Oct Monday)

Monday morning, after visiting David's electronic shop on the Harvard campus on Saturday and acquiring the fully loaded BlackBerry that David ordered, Christa presents her new electronic communication device to David for inspection and instruction for use. David removes the Blackberry from the shipping container, activates the unit and then carefully demonstrates all the features. Christa is duly impressed with David's careful and gentle instructions regarding the BlackBerry's capabilities. David accesses his own email on his Blackberry and activates Christa's email account on Google using Gmail. David quips that he will test her Gmail reception and quickly texts a brief message from his BlackBerry to Christa's BlackBerry.

Christa presses the receive icon on the BlackBerry screen and smiles as she opens and views David's short message on her screen that reads,

"You are especially lovely today Christa, and your presence always brightens my Hospital Suite…but that is true every day in this stolid hospital ward when you grace my hospital suite with your beauty and attendance."

Christa quickly texts a response to David that reads,

"Thank you, my handsome and brilliant young ward, for the compliment. You know there is no other place I would rather be than with you this day."

David then demonstrates the GPS and its capacity to identify and locate shops and addresses of interest. She can now identify and locate shops and grocery stores near David's apartment and even compare prices for selected items. David then probes his Hospital Suite with the camera on the phone and zooms in to identify objects of interest. Christa then tests the BlackBerry camera and confirms her ability to access and transmit visual images. She smiles and states she will use the BlackBerry camera when she goes to the Hospital galley for David's meals, so he can select his choice. Maybe then, David will cease his complaints about the hospital food she brings to feed him.

Christa and David will find that their BlackBerrys are very effective for communication and their use will greatly enhance their companionship and time with one another. Christa soon becomes very proficient with her BlackBerry and she fully uses this device to intensify her interactions with David. They spend considerable time communicating in the evening after Christa has returned to the apartment. Christa often leaves the BlackBerry line open when she is in the apartment, so they can freely communicate, and David can easily request the research materials he needs upon her return to the hospital. As she planned, Christa realizes that she is now an indispensable source for supporting David's research and providing immediate response to his needs. David often tasks Christa to locate a piece of information from his library and transmit the information to David either verbally or as text via their BlackBerrys.

David and Christa BlackBerry communication (26 Oct Tuesday)

With quasi-continuous exchange of personal, scientific and religious dialog between the devout Mormon nurse and the brilliant cosmological physicist, the inevitable bonding is occurring. Now, using their mutual BlackBerrys, Christa calls David upon her arrival in the apartment from the Hospital as David has requested. David has expressed his need for confirmation of Christa's safe return to the apartment, so she affirms her arrival soon after entering the apartment. It delights Christa that David is very concerned for her safety and well-being. It is evidence to Christa that David loves her, and they are on the path to marriage and David's embrace of the Gospel.

They soon develop a routine that includes Christa opening the mail they both receive and discussing any necessary financial affairs and responses to correspondence. Christa sets the BlackBerry on the kitchen table while Christa changes from her nursing uniform to her causal evening clothes. Christa goes about her routine household duties with the BlackBerry camera active. They converse and interact together throughout the evening with their individual BlackBerrys. David realizes that these electronic marvels provide vicarious familiarity and intimacy for the exchange of their emotions as well as their thoughts and feelings. David realizes that the only improvement over this means of interaction would be if David were present with Christa in the apartment and they were married.

Christa will often, after her evening meal and household duties, settle into a comfortable position on the recliner in her room and read from the Scriptures for David. David greatly enjoys Christa's soft and pleasant elocution and their intimate discussions about the Gospel. David informs Christa that their exchange is almost as if she were by his side in his Hospital Suite reading to him as a mother would to her child. David opines that he missed this nurturing rite in his childhood.

David is also able to visually observe, through the BlackBerry camera, his apartment, Christa's housekeeping talents, his scientific research materials, and, most important for David, to observe Christa. He can identify and direct Christa where to locate his scientific materials among his papers and books that he requires to support his research. However, the major use of their BlackBerrys is their extensive exchange of communication and interaction with emphasis upon the Gospel and David's growing testimony. This electronic conveyer of messages as electromagnetic signals become their major means for exchanging observations, feelings, ideas, activities and aspirations outside David's Hospital Suite. Indeed, even after eight hours or more of the normal nursing schedule within David's Hospital Suite during the day, they continue to interact throughout the evening often until both retire. Their mutually paired and connected BlackBerrys become their principal medium for communication and progressive understanding and deepening love for each other. David and Christa do not have the

93

constraints and formality that prevails in David's Hospital Suite when they are alone and joined with their respective BlackBerrys.

Christa informs and displays for David, employing the BlackBerry, her homemaking skills and her personal touch in organizing the apartment. She displays her room and her feminine touch to David in domestic homemaking. Christa's bookshelf holds copies of the standard Mormon Scriptures, the Holy Bible, the Book of Mormon, the Doctrine and Covenants, and the Pearl of Great Price. The shelf also holds a quadruple combination of all four Scriptures and monthly copies of the Church Magazine, the Ensign. On the wall above her bed, a picture of the Salt Lake Temple is prominently displayed. On the opposite wall is a singular, large portrait of the "Christus" statue that is in the visitor's center on Temple Square in Salt Lake City. David has observed this magnificent, white marble sculpture frequently when Christa has spanned the room with her BlackBerry. David acknowledges that the representation is accurate, and it does portray an early Jewish Rabi who would be a member of the House of David.

In David's Study is his desk, clear and unencumbered from its earlier, disorganized state. The desk now has a pen and pad carefully placed next to his other Apple laptop computer. Furthermore, all his books are carefully aligned by subject matter along the bookshelf that reaches from the floor to the ceiling. David is very impressed with her logical assortment of his literary titles. To his great surprise, Christa has located some books that he had earlier lost or misplaced. Viewing the shelves through Christa's BlackBerry, David recognizes that his books are arranged as a science librarian would arrange for the shelves at Harvard's Widener Library. Indeed, he can now readily select what he needs through the BlackBerry and Christa can deliver the selection to him the next weekday. With Christa panning the BlackBerry camera in David's bedroom, he observes that his room is now immaculate and carefully ordered. His clothes closet is well organized with shirts, pants, and shoes are smartly sorted in an appropriate manner. His shoes are now polished, and shoetrees are evident in each pair of shoes. Even his old athletic shoes have been washed and are properly stowed. David surmises that his athletic shoes have now lost their strong scent of perspiration, although even his highly developed sense of smell cannot detect the electromagnetic signal for the odor of perspiration delivered through the BlackBerry, that is, if there is such a physical signal.

BlackBerry impact on Christa (31 Oct Sunday)

David, in witnessing this display of his apartment delivered to him by Christa, ponders with silent amusement what his life would be if Christa were his wife. The thought delights and tantalizes him. Besides her beauty, feminine form, and exceptional and provocative mind, she is an accomplished homemaker, manager, and an indispensable figure in his life now. If he could only taste the attractive dinners she prepares for herself in the evening, he surmises that he could evaluate her culinary skills. However, there are limits to what a BlackBerry's electromagnetic communication capabilities can transmit.

On Sunday, Christa takes her BlackBerry to her Mormon Ward and allows David to listen to church services when appropriate. After Church services, Christa introduces David with her BlackBerry to her Ward members and leaders. They can also observe David visually and orally and become acquainted with this scientific giant. Christa's friends privately chide her regarding her unusual mode of courtship, but they reflect her matrimonial plans. Christa's female companions assert that her ubiquitous, visual and oral communication with David means she is connubially joined to David through her BlackBerry. They frivolously label her relationship with David as a BlackBerry marriage, only lacking physical intimacy.

Christa's Journal entry (31 Oct Sunday)

On Sunday, after attending the three-hour church service, preparing her simple meal, and loading the automatic dishwasher for its infrequent weekly operation, Christa calls David on the BlackBerry, and she pans the apartment with the internal camera. David is impressed with the improvements and changes that Christa has wrought within the apartment.

Christa informs David that she must make her weekly telephone call to her mother in Utah and requests that David join the conversation via their BlackBerrys. Christa states that her mother wants to become personally acquainted with David since Christa has spoken frequently regarding her solitary care for him at the hospital. Christa then reinforces her request saying,

"My mother is anxious to hear what an endowed, Harvard Physics Professor says and does."

David responds with self-deprecating parody,

"Christa, your mother may be disappointed when she actually hears me speak on the telephone. I may not meet her maternal expectations for you, and I do not want her to tell you to abandon me as your patient. I have invested too much time and interest in you to lose you now."

Christa assures David that her mother will be as enamored and impressed with David as Christa is with David. Christa states she has been very complimentary regarding her description of David to her mother.

The mutual, three-way telephone conversation with David, Christa and her mother begins and proves to be a congenial and poignant communication. They discuss many events and interests including Christa's childhood and her father, David's family and his background, and a brief account of David's pursuits in M-Theory. After a lengthy and convivial exchange of words and shared feelings, David expresses his deep appreciation for Christa and her tender and competent care for him. He is very dependent upon her and the important role she now occupies in his life. As David closes his participation in the telephone exchange, he wishes Christa's mother a pleasant evening and he ends his conservation with the two women.

Christa continues the dialogue with her mother and the mother and daughter offer their candid assessments of David Steinmann. It is apparent to Christa's mother as Christa concludes her long-distance telephone call with her daughter that Christa deeply loves David. After expressing her love and closes her telephone call, Christa opens her journal and records the past events for the week.

Christa's weekly Personal Journal for 25-31 Oct 2010

I am now fully settled into David's apartment and David's apartment manager and all his apartment neighbors are friendly and accept me as their new neighbor. An older woman who is a widow and lives alone in David's apartment complex has invited me to have dinner with her whenever I can. She is very lonesome living alone, and I will try to visit her often.

I occupy Rebecca's room and have boxed, labeled and stored Rebecca's personal articles awaiting instructions from David and his mother for disposition of these items.

The apartment is large and comfortable but will require my continued attention and effort to make it a home – perhaps a future residence for David and me if my plans for us transpire. I am certain that my mother is now aware of the depth of my need and love for David.

In closing my journal entry, I am resolved and confident that David will be the man in my life who will become my husband and eternal companion. Bless me to that end, Eternal Father, if that is Thy will also.

CHAPTER 7 (25 Oct-31 Oct)

David's declaration of agnosticism (25Oct)

During one of their many frequent and fervent discussions regarding science and religion, David admits that he is agnostic and has no strong religious persuasion or commitment to religion. David acknowledges that "Intelligent Design" advocates pose significant challenges to Darwin's supposition for the evolutionary development of life on the earth based only upon random processes and statistical outcomes. Darwin's "Origin of Species" must begin with an active living cell outfitted with the necessary genomes required for subsequent reproduction. The complexity of the genome with its 3 billion base pairs forming the 1 billion codons that comprise and produce the 20 amino acids in the human genome could not be a simple product of random assemblage of its chemical elements. Even the 13.8 billion years span since the "Big Bang" is insufficient time for random assemblage of a single large protein. David believes that the primitive genome for life was probably introduced via a parallel universe.

However, David's basis for agnosticism stems from his early childhood. His stern, opinionated, Jewish mother, who is a rabbinical Jew and tolerated no deviation from her rigid Jewish orthodoxy, demanded strict conformance to the edicts of the Jewish Torah. David tells Christa,

"At the age of 13, I dutifully performed my 'Bar Mitzvah' at the Jewish Western Wall in Jerusalem. I read from the Torah in Hebrew as is expected and required for young, orthodox, Jewish males. However, I performed this ritual only to satisfy my mother and please the rabbinical relatives on her Jewish line. Because of my mother's dominant, intransigency regarding her orthodox faith; I have rejected my childhood religion. Most of my scientific colleagues are also either agnostics or atheists. They do not believe that a 'God or Supreme Being' is evident or necessary in science, in the universe, or even in human affairs."

David's declaration of agnosticism deeply offends Christa and she responds with the firm rebuttal,

"David, I am greatly disappointed in your words. To not know of God or even search for God is a great disappointment for me. David, I earnestly pray nightly to our Heavenly Father for your full recovery, and God may not grant my petitions if you reject God."

Christa's anguish is evident to David as Christa continues solemnly,

"Even if David does not know God, God knows David and loves David. God loves you David, and I love you, and I pray for you constantly."

With those emotional words and her open declaration of love for David, a somber and subdued Christa quietly, but hastily, leaves his Hospital Suite. During her absence, David ponders her declaration and the impact that his words have had upon her. He realizes that his opinions in religious matters are very important for Christa. Their relationship is becoming very intimate and tender, and David's religious commitment is obviously very important to Christa. He has offended her, and she is far too important and essential in his life now to be offended.

David and Christa will exchange religion and science (25 Oct)

While Christa is absent from David's Hospital Suite and she sits at the nursing station for the duty nurse, she is deeply disturbed by David's admission of unbelief in God. She realizes her love for David is intensifying, but she cannot accept and bring him in her life unless he believes in God the Father and His son, Jesus Christ. She must have a companion who believes in Christ, accepts His principals for living, and obeys His commandments including baptism and confirmation. Pondering this dilemma for her future relationship with David, Christa methodically suppresses her disappointment in David's avowed agnosticism and returns to David's Hospital Suite. She resumes her necessary nursing services for David, but without verbal interaction and her usual playful banter with him.

David, observing Christa's aloof demeanor, thoughtfully ponders his response to Christa. David then offers this provocative compromise.

"Christa, I realize my admission of agnosticism disturbed you and I sincerely apologize. My statement was made without forethought. After prudent deliberation, I have a compromise to offer you for your consideration. I hope you will consider my sincere proposal with sympathetic consideration."

David carefully selects his words as he declares to an undecided Christa,

"I will attempt to explain the realm of physics and cosmology that I know and occupy, if you will acquaint me with your God....your Christian, Mormon God...Jesus Christ. Christa, I know that the Latin root for Christ is Christus and the Greek root is Christós. Your name, Christa, derives from both roots and signifies a follower or Disciple of Christ. I recognize and honor your deep religious commitment to Christ and His Gospel. I am somewhat familiar with Christian theology and I understand that the sacred writings in the book of Saint John in the New Testament are an excellent beginning for gaining an understanding and testimony of Christ according to my Christian colleagues at Harvard."

David then declares with solemn words,

"Christa, will you read to me and discuss the scriptural verses from the Book of Saint John and other scriptures in the New Testament and your Mormon Scripture, the Book of Mormon; I believe you call your scripture?"

David, still seeking to secure an exonerating façade on Christa's countenance, then suggests with subdued and playful wit,

"If you will select and discuss the Christian 'Gospel of Christ' according to your Mormon beliefs with me, then I will relate to you my beliefs in science, let me call it the 'Gospel of Science' according to David ben-Steinmann."

David observes rising amelioration in Christa's demeanor, so he continues with his attempt to secure her participation with him regarding religion and science by declaring,

"I will seek to understand your beliefs and feelings if you will listen to my ideas and physical theories and ponder them. Perhaps, if I can clearly and simply explain to you my claims and evidence for M-Theory, then others may also be persuaded that M-Theory is worthwhile and a valid venture in science."

David persists in his effort to mollify Christa's disappointment regarding his declaration of agnosticism.

"I know that you firmly believe in the existence of your God...Jesus, and His Heavenly Father. They are essential beings in your life and heart and my growing love for you impels me to acquire a recognition and hopefully a testimony of Christ and His Gospel. "

David then pauses and proffers his final proposal to pacify Christa.

"Perhaps your religious faith and scriptures will impart religious faith for me and may offer support for my research. My NDE has provoked considerable deliberation on my part and the meaning of the events that I experienced. My NDE may have a religious and spiritual basis and truth that Mormonism can address and explain. Perhaps, I too may then become a Disciple of Christ."

To David's relief and joy, Christa approaches David's bed and with a warm smile, she gently rests her hand on his chest and declares,

"David, I accept your apology and fully support your proposal. I desperately want you to know of God the Father and His Son, Jesus Christ. These Beings are real and are central in my life. I want to share them with you because they are essential for my happiness and our future relationship. Yes, David, I will joyfully share my knowledge and testimony of Christ with you; and you can help me to understand and appreciate your immense knowledge in science. Perhaps, Christian and Mormon Scriptures will provide important answers to your NDE while also increasing your faith in Christ and bringing me great satisfaction and joy."

Christa consummates her offer by declaring,

"David, all the essential Mormon scriptures are available for download on your I-Pad. The Hospital has Wi-Fi throughout the Hospital and I can download onto your I-Pad all the relevant scriptures including the Old and New Testament, the Book of Mormon, the Doctrine and Covenants, the Pearl of Great Price, and many of the other Mormon Church works. Most important, these scripture and writings are supported by topical guides and indices that allow you to scan any of these works and read the referenced scripture. Remember, I referred to the Veil that Mormon's believe separates our mortal world from the Kingdom of Heaven. The Veil is also an important component of the sacred ordinances we perform in our Temples. Significantly, this Mormon Veil may have relationship to the membrane that you assert in M-Theory that separates our four-dimensional spacetime universe from a parallel universe. Perhaps the parallel universe you conjecture in M-Theory is the Domain that God and His Hosts occupy. During your NDE, you may have been a brief visitor in the Kingdom of Heaven. Significantly, Mormons also believe that God and Christ can move freely through time and space without the constraints that mortals encounter. You may have briefly experienced these properties in your NDE. Also, the past, present and future time are open to the servants in God's Domain and these servants can interact with mortals as directed by our Heavenly Father, Elohim, and His Son, Jesus Christ."

"We believe that the original Church of Jesus Christ established by the Savior was reestablished through a sacred visitation by God, the Father and His Son, Jesus Christ to Joseph Smith, the Mormon Prophet." Later, an earlier inhabitant of the Americas appeared as the resurrected angel, Moroni, to Joseph Smith. This angel was the mortal author of a record inscribed upon plates documenting an ancient people who lived in the

Americas and were visited by Christ after His crucifixion and resurrection. These plates contain the Book of Mormon that Mormons believe is another testimony of Jesus Christ."

Christa concludes her extended assessment with obvious enthusiasm and optimism as she declares.

"David, as you can tell, I am excited with your offer for a mutual exchange of beliefs and experiences. I believe that this exchange will greatly enhance our interactions and understanding with one another and my fervent hopes and prayers for our future relationship."

With these mutual agreements, David and Christa begin their sustained interchange of religious and scientific beliefs and principles. They will now devote much of their interactions and time with one another discussing and exploring religion and science and their understanding and evidence for both.

Later, David asks Christa to purchase two large screen Kindles for ease in perusing these Scriptures. Then, with each of them with these convenient resources, they can jointly access and discuss scriptural references. Christa purchases large screen Kindles for their mutual use and downloads all the Mormon Scriptures onto these devices for their mutual use.

This enhanced access to the Scriptures greatly benefits David and Christa as they explore the Gospel. Christa and David use their kindles in their extensive investigation into the Gospel of Jesus Christ. Upon Christa's suggestion, David prepares for their first religious exchange by carefully reading the record of the Saint John in the New Testament. This inspired scripture serves as the beginning record for their mutual exploration and discussion of the Gospel.

David and Christa religious discussions in St. John

Christa initiates her intense and intimate Gospel discussions with David by beginning with the first chapter of Saint John in the New Testament. Christa testifies to David,

"The Book of John in the New Testament is a particularly cogent and effective Scripture to begin our discussions about the mission of the Savior of the World, your Messiah, Jesus Christ. This First Chapter in St. John declares that Christ is the Word of God. He created all things, was made flesh, and came to the earth to expiate the sins of humanity through His Atonement. Christ's Atonement gained resurrection and immortality for all mortals and will allow humanity to return to the Father of their Spirits, Elohim or God the Father. I believe that Jews refer to God the Father as Elohim, and so do Mormons."

David then reads the following verses for Christa regarding Christ and His role in the Universe we inhabit.

John 1
1. *In the beginning was the Word, and the Word was with God, and the Word was God.*
2. *The same was in the beginning with God.*
3. *All things were made by him; and without him was not anything made that was made.*
4. *In him was life; and the life was the light of men.*
5. *And the light shineth in darkness; and the darkness comprehended it not.*
9. *That was the true Light, which lighteth every man that cometh into the world.*
10. *He was in the world, and the world was made by him, and the world knew him not.*
11. *He came unto his own, and his own received him not.*
12. *But as many as received him, to them gave he power to become the sons of God, even to them that believe on his name:*

As David completes reading, Christa declares to David,

"Jesus Christ is not only the Messiah sought by your people, the Jews throughout their history, but He is the only Begotten Son of God the Father, Elohim. Christ supervised the establishment of the Earth and other planets where human life is found. It was Christ, under the direction of His spiritual and physical Father, Elohim, who fashioned the genome based upon DNA as the basic building structure for all living things. As I learned in nursing school, all life forms contain one or two genomes. Most advanced flora and fauna including humans contain two genome types while fungi, algae and bacteria contain only one genome. DNA or deoxyribonucleic acid is the double helix, molecular chain that encodes all the essential genetic information within its molecular structure required to generate and sustain animal life throughout the universe we inhabit."

Christa continues,

"The two intertwined nucleotide chains in DNA are oriented in opposite directions and the nucleotide base pairs are joined by hydrogen bonds that can be severed and rebounded. During DNA replication, these chains separate, and the base pairs serve as unique templates for making identical daughter molecules of DNA. The functional units of segments of the DNA are the genes that can be copied to make RNA (ribonucleic acid). RNA can then be translated into the amino acid sequences to make proteins. Proteins are the organic molecules that are fundamental components of all living cells. Proteins consist of 20 essential amino acids linked in a genetically controlled linear sequence forming long polypeptide chains that form collagen, hemoglobin, antibodies and enzymes. These macro molecules provide the building blocks for all living matter."

David smiles at Christa and complements her on the quality of the nursing education she received at BYU. David then comments that on a statistical basis, the number of ways that 20 amino acids can be randomly distributed to form a simple single protein with only 100 assigned amino acids is equal to the number of amino acids, 20, raised to the power of 100 for the number of amino acids in this simple protein. As an equation David concludes that number of possible arrangements is 20^{100} or about 10^{130}.

The value of this estimate is staggering since cosmologists believe that the total number of atoms in our universe, as equivalent hydrogen atoms, is about $\sim 10^{80}$. So, the probability of forming a single protein in any possible multiverse similar to ours, implies that to form this simple protein would require the existence of about 10 ^{50} universes like ours.

Christa, delighted by David's calculations and his recognition of her knowledge of microbiology, continues,

"David, that is a compelling argument for 'Intelligent Design" or God. The genetic code for DNA found in all living organisms is comprised of the two nucleotide pairs, adenine and thymine, abbreviated as (A-T), and cytosine and guanine, abbreviated as (C-G). These nucleotides are arranged along the double helix spiral backbone supporting these base pairs. These four nucleotides program all life forms from simple algae to complex human germ cells. The haploid human genome with 23 chromosomes is estimated to have about 3 billion base pairs and contain 20,000 to 25,000 distinct genes that are critical for human transcription."

Christa then expresses her opinion regarding the non-critical genes,

"I personally believe that some of the noncritical genes were among those that were altered when Adam and Eve became mortal in the Garden of Eden. It is these altered or corrupted genes that result in human mortality and death. When we are resurrected, I believe that these corrupted genes will be corrected, and we will become immortal, resurrected beings. Regardless, however we achieve immortality, this glorious, incomparable transition was made possible by the Atonement of Jesus Christ."

Christa then confides to David, that the choice of the letters for the base pairs is very significant and profound for her. Christa associates the unique C-G nucleotide pair in DNA with "C" for Christ and "G" for God the Father or Elohim. This base pair symbolizes the essential union of the Father and the Son and their universal eminence and singular importance for all life forms in our universe including David's multiverses. Furthermore, for the nucleotide pair A-T, Christa associates "A" for the Atonement wrought by Christ that gained resurrection and immortality for all mortals. Christ's Atonement will restore the original eternality of the genome for humanity.

Christa continues her explanation on the base pairs in DNA,

"The acronym 'T' for the nucleotide thymine conveys the terms, Testimony and Truth. Christ declared in St John, Chapter 8, and verse 32. *'And ye shall know the truth, and the truth shall make you free'.'*"

Christa then declares,

"The 'truth' is that Christ's Atonement has made all humans free from the fall of Adam and eternal death and all will participate in the resurrection and obtain immortality. The Spirit of Truth is the Holy Ghost who confirms our testimony in Christ and is the medium by which mortals gain access to the Kingdom of Heaven. We gain access to our Heavenly Father through the Holy Ghost by gaining a testimony of Christ and becoming a Disciple of Christ."

"Christ and His Father have life within themselves or eternal life and their genomes, which are immortal, provide them with eternal life. There are many Scriptural passages attesting to the fact that Christ permitted His physical death to occur, although He had eternal life within Himself because He was the only Begotten Son of the Father in the flesh and thus possessed these immortal genomes from His Father that are essential for eternal life."

"Adam and Eve were also originally bestowed with these eternal genomes, but their fall in the Garden of Eden brought corruption to their immortal genomes and eventual death to their mortal bodies. All progenitors of Adam and Eve have these mortal genomes and are subject to death of the physical body. But the Atonement of Jesus Christ will restore these mortal genomes to their eternal state, and everyone born on Earth will be resurrected and become immortal."

David and Christa discuss the importance of baptism for discipleship with Christ

Christa then tells David that baptism is the essential first step in Discipleship with Christ after one has gained faith and belief in Christ and repented of his or her mortal sins, Christa confirms this by reading the following narrative regarding Christ's baptism by John the Baptist.

> **John 1**
> 29. *The next day John seeth Jesus coming unto him, and saith, Behold the Lamb of God, which taketh away the sin of the world.*
> 32. *And John bare record, saying, I saw the Spirit descending from heaven like a dove, and it abode upon him.*
> 33. *And I knew him not: but he that sent me to baptize with water, the same said unto me, upon whom thou shalt see the Spirit descending, and remaining on him, the same is he which baptizeth with the Holy Ghost.*
> 34. *And I saw, and bare record that this is the Son of God.*

Christa solemnly testifies to David the essential nature of Baptism by saying,

"Baptism is significant both physically as well as spiritually. Spiritually, baptism confirms the Holy Ghost upon the recipient and the Holy Ghost is the medium by which the baptized person can interact with both the Father and Son. Furthermore, as Christ was physically baptized by John and dwelt as a mortal on the earth, some of Christ's immortal genomes were released from his body into the waters and air that now cover the Earth."

Christa and David discuss the reality and significance of this observation that we are washed by the blood shed by the Christ. Christa asks David to estimate the number of atoms within the air and water that passed through Christ's body during his thirty-three years of mortality on the earth.

David, with Christa's participation, estimates the distribution of Christ's mortal fluids throughout the earth and quickly obtains the following results that he documents on his I-pad as follows.

David's calculation

Claim: During His mortal life and crucifixion, Christ's blood and body containing his mortal and immortal DNA was dispersed throughout Earth's biosphere. Earth's waters, soils, and atmosphere now hold very small but measurable amounts of the residue of His blood and breath. During 33-year mortal life of Christ, His unique DNA permeated Earth's biosphere.

Calculation: Assuming human properties and functions for 70 kg male living for 33 years, David calculates Respiration rate at 2 cubic-meters/day x 365 days/yr x 33 yrs

= 24,000 cubic-meters (m3 of air)

Fluid exchange rate 2 liters/day x 365 day/yr x 33 years

= 24,000 liters

For respired air and fluid (water) exchange, we have:

Molecules/m3 (air)

= 0.6E24 molecules/gmw/(28 g/gmw) x 1E-3 g/cc x 1E6 cc/m3

= 2E25 molecules/m3

Molecules/liter (fluid as water)

= 0.6E24 molecules/gm./(18 g/gmw) x 1 g/cc x 1E3 cc/liter

= 3E25 molecules/liter

Christ's body (flesh) and blood (fluids) that exchanged with Earth's biosphere yielded following molecular quantities:

= 2E25 molecules/m3 (air) x 24,000 m3 (released to Earth's atmosphere)

+ 3E25 molecules (fluid)/liter x 24,000 liters

= 1E30 molecules   Equation (1)

To approximate total water volume on Earth, assume Earth's total water mass is distributed uniformly over the surface.

Earth's radius R is about 4000 miles or 6.4E6 meters (m) with an average depth D of 1 mile or 1610 meters (m) so

Total water (m3) on Earth = $4 \pi R^2 D = 4 \pi (6.4E6 \text{ m})^2$ x 1610 m

= 8.3E17 m3 = 8.3 E23 cc   Equation (2)

Number of molecules of Christ's bodily fluids (air, water, and blood) per cc of Earth's water and air is given by

= 1E30 molecules / [8.3 E23 cc]   Equation (1) / Equation (2)

= 1.2E6 molecules/cc or about 1 million molecules per cc or gram of water

David underlines his conclusion:

About 1 million molecules that passed through Christ's body during His mortality on earth are present in every milliliter of water on the Earth.

David informs Christa that his calculation reveals about one million atoms that transited Christ's body during His mortality. Christa and David discuss the nature of the fluids (both air and water) that passed through Christ during His thirty-three-year mortal sojourn. These atoms have now dispersed throughout the Earth's atmosphere and water and now contain about one million atoms in each gram of water and air. Water and air are the essential fluids to support life found anywhere in the universe based on the genome for life. Thus, the air and water on the Earth contain atoms that passed through the body of Christ and very small amounts of Christ's genomes have permeated this water and air throughout the earth.

Christa then declares that the scriptural statement "we are washed by blood of Christ" is scripturally accurate and factually true. We are literally cleansed with blood of Christ as we breathe, eat, bathe and are baptized and partake of the Sacrament. It is Christ' sacred blood that contained His immortal DNA that now insures human resurrection and immortality through the bestowal of the Holy Ghost.

Christ called Apostles to declare his Gospel throughout the world.

Christa and David discuss the organization that Christ established during His mortal ministry to declare His Gospel throughout the Earth. Christ began by gathering His apostles to declare and administer His Gospel on the Earth. He told Philip on being called as an Apostle that Phillip would witness the heavens open and the angels of God, His Father, ascending and descending upon Christ, the Son of Man. To support the calling of the Apostles, Christa reads from John the following.

John 1

45 *Philip findeth Nathanael, and saith unto him, we have found him, of whom Moses in the law, and the prophets, did write, Jesus of Nazareth, the son of Joseph.*

47 *Jesus saw Nathanael coming to him, and saith of him, Behold an Israelite indeed, in whom is no guile!*

48 *Nathanael saith unto him, Whence knowest thou me? Jesus answered and said unto him, Before that Philip called thee, when thou wast under the fig tree, I saw thee.*

49 *Nathanael answered and saith unto him, Rabbi, thou art the Son of God; thou art the King of Israel.*

50 *Jesus answered and said unto him, Because I said unto thee, I saw thee under the fig tree, believest thou? thou shalt see greater things than these.*

51 *And he saith unto him, Verily, verily, I say unto you, Hereafter ye shall see heaven open, and the angels of God ascending and descending upon the Son of Man.*

Christa testifies to David that the presence of Apostles is essential for the establishment and sustainment of Christ's Church on the Earth. Christa then humbly testifies that the Church of Jesus Christ of Latter Day Saints is the only Christian Church on the earth today that has Twelve Apostles who are ordained of God and serve as special witnesses of Jesus Christ throughout the Earth.

Christ's miracles demonstrate He is the Only Begotten Son of the God

Christa declares the many miracles performed by Christ during his ministry on the Earth. Christ, as the Son of God, can perform miracles that are the province and privilege of deity. Furthermore, since Christ was responsible for fashioning the genome for life, He could readily alter the chemistry of water to produce wine as witnessed in His first miracle ... the conversion of water to wine. Christa reads John 2 where Jesus turns water into wine during a wedding feast in Cana. Observe that the mother of Jesus was aware that her Son was divine, but Christ was reluctant to display His divine powers...nevertheless, to please His mother, He performed the miracle declared in John's following scripture.

John 2
1. *And the third day there was a marriage in Cana of Galilee; and the mother of Jesus was there:*
3. *And when they wanted wine, the mother of Jesus saith unto him, They have no wine.*
4. *Jesus saith unto her, Woman, what have I to do with thee? mine hour is not yet come.*
5. *His mother saith unto the servants, Whatsoever he saith unto you, do it.*
6. *And there were set there six waterpots of stone, after the manner of the purifying of the Jews, containing two or three firkins apiece.*
7. *Jesus saith unto them, Fill the waterpots with water. And they filled them up to the brim.*
8. *And he saith unto them, Draw out now, and bear unto the governor of the feast. And they bare it.*
9. *When the ruler of the feast had tasted the water that was made wine, and knew not whence it was: (but the servants which drew the water knew;) the governor of the feast called the bridegroom,*
10. *And saith unto him, Every man at the beginning doth set forth good wine; and when men have well drunk, then that which is worse: but thou hast kept the good wine until now.*

Christa explains the role that Temples represent at the time of Christ

Christa states that Christ's mission on the earth was to perform the works that His Father, Elohim had required of Him during His mortal mission on Earth. In Chapter 2 of John, Christ declared that the Temple in Jerusalem was His Father's House. Because of the sanctity of the Temple, the mortal Christ was offended that the Temple in Jerusalem, His Father's House, was being desecrated by mercenary merchants. Today, sacred Temples numbering more than a hundred throughout the earth have been built or in preparation by Mormons. These sacred Temples are vital to the salvation of all earthly mortals and serve as the temporary abode of Christ and heavenly beings when they visit the earth.

Christa also declares that Christ informed the Jews that if they destroyed His mortal body (His mortal temple), He would raise up His body again after three days. Christ was foretelling of His resurrection from the dead in the following scripture.

John 2:

13 *And the Jews' passover was at hand, and Jesus went up to Jerusalem,*

14 *And found in the temple those that sold oxen and sheep and doves, and the changers of money sitting:*

15 *And when he had made a scourge of small cords, he drove them all out of the temple, and the sheep, and the oxen; and poured out the changers' money, and overthrew the tables;*

16 *And said unto them that sold doves, Take these things hence; make not my Father's house an house of merchandise.*

18 *Then answered the Jews and said unto him, What sign shewest thou unto us, seeing that thou doest these things?*

19 *Jesus answered and said unto them, Destroy this temple, and in three days I will raise it up.*

20 *Then said the Jews, Forty and six years was this temple in building, and wilt thou rear it up in three days?*

21 *But he spake of the temple of his body.*

Christ tells His role as the Son of God to Nicodemus, a ruler of the Jews

David and Christa then discuss the interaction that Jesus had with the Pharisees and Sadducees. Jewish tradition held that a Messiah would come to the Jews and there was belief among some Jews that Christ was the Messiah. Christa cites the following verses from John 3 to David regarding Christ's visit with Nicodemus, a ruler of the Jews who was probably a Pharisee. Nicodemus was a devout Jew and he recognized Christ's divinity and acknowledged that Jesus was a Rabbi sent by God...and possibly the Jewish Messiah. Christa then reads the following verses from John, Chapter 3 to support her declaration.

John 3

1   There was a man of the Pharisees, named Nicodemus, a ruler of the Jews:
2   The same came to Jesus by night, and said unto him, Rabbi, we know that thou art a teacher come from God: for no man can do these miracles that thou doest, except God be with him.
3   Jesus answered and said unto him, Verily, verily, I say unto thee, Except a man be born again, he cannot see the Kingdom of Heaven.
4   Nicodemus saith unto him, How can a man be born when he is old? can he enter the second time into his mother's womb, and be born?
5   Jesus answered, Verily, verily, I say unto thee, Except a man be born of water and of the Spirit, he cannot enter into the Kingdom of Heaven.
6   That which is born of the flesh is flesh; and that which is born of the Spirit is spirit.
7   Marvel not that I said unto thee, Ye must be born again.

Christa tells David that this verse in John declares the essential act that man must perform to receive the Kingdom of Heaven, the abode of God the Father. That act is Baptism or to be reborn of water and receive of the Holy Ghost, the Spirit Member of the Godhead. Without these saving ordinances of "being born again," man cannot dwell with his Heavenly Father in the Kingdom of Heaven. Christa states that the precise manner and source for the bestowal of the Holy Ghost is unknown to man as declared in these verses. The Holy Ghost is the Spirit member of the Godhead and has the power and assignment to communicate with man across the veil that separates man and God. Christa then expresses her belief that it was through the Holy Ghost that David received and experienced his NDE. However, the precise mortal way the Holy Ghost is bestowed and operates upon man is unknown. Observe that Nicodemus was unaware of how the Holy Ghost visited man and Christ's response to Nicodemus was that the actual manifestation and bestowal of the Holy Ghost is unknown to mortals.

John 3

8 *The wind bloweth where it listeth, and thou hearest the sound thereof, but canst not tell whence it cometh, and whither it goeth: so is every one that is born of the Spirit.*

9 *Nicodemus answered and said unto him, How can these things be?*

10 *Jesus answered and said unto him, Art thou a master of Israel, and knowest not these things?*

11 *Verily, verily, I say unto thee, We speak that we do know, and testify that we have seen; and ye receive not our witness.*

12 *if I have told you earthly things, and ye believe not, how shall ye believe, if I tell you of heavenly things?*

13 *And no man hath ascended up to heaven, but he that came down from heaven, even the Son of man which is in heaven.*

14 *And as Moses lifted up the serpent in the wilderness, even so must the Son of Man be lifted up:*

15 *That whosoever believeth in him should not perish, but have eternal life.*

16 *For God so loved the world, that he gave his only begotten Son, that whosoever believeth in him should not perish, but have everlasting life.*

17 *For God sent not his Son into the world to condemn the world; but that the world through him might be saved.*

18 *He that believeth on him is not condemned: but he that believeth not is condemned already, because he hath not believed in the name of the only begotten Son of God.*

35 *The Father loveth the Son, and hath given all things into his hand.*

36 *He that believeth on the Son hath everlasting life: and he that believeth not the Son shall not see life; but the wrath of God abideth on him.*

After reading John 3, Christa testifies to David that,

"Christ declared to Nicodemus that He, Christ, was the Messiah and we must believe in Him and receive Christ to receive the Holy Ghost and be spiritually born again to receive eternal life and residence in the Kingdom of Heaven with Elohim. It is through the power and agency of the Holy Ghost that man is spiritually reborn and gains access to the Kingdom of Heaven. Furthermore, Jesus testified to Nicodemus that His Father, Elohim, had sent Jesus as His only begotten son in the flesh to secure eternal life for man.

Man would then receive the immortal genomes possessed by Christ to accomplish this."

Christa then solemnly repeats the marvelous promise made by God the Father in the 16th verse of John 3 that Jesus Christ, the true Messiah sought by the Jews, offers eternal life to all those who believe that He is the only begotten Son of God in the Flesh.

> *"For God so loved the world, that he gave his only begotten Son,*
> *that whosoever believeth in him should not perish, but have everlasting life"*

Christa testifies that Christ declared his Messiahship to a Gentile.

David and Christa then read John 4 together. Christa says that Christ also testified to a woman of Samaria, a non-Jew, that He was the Messiah, the Son of God, sent from God, the Father. To please the Father, all mortals must accept and worship Christ. It is the Holy Ghost who confirms that Christ is the true and only path to return to the Father. This is essential for all mankind, as Christ told the woman, that Christ came to earth for all mortal humans as well as the Jews. Christ declared to the Samaritan woman, what all mortals must do to receive forgiveness, immortality and exaltation in the Kingdom of Heaven.

John 4

6   *Now Jacob's well was there. Jesus therefore, being wearied with his journey, sat thus on the well: and it was about the sixth hour.*

7   *There cometh a woman of Samaria to draw water: Jesus saith unto her, Give me to drink.*

9   *Then saith the woman of Samaria unto him, How is it that thou, being a Jew, askest drink of me, which am a woman of Samaria? For the Jews have no dealings with the Samaritans.*

10   *Jesus answered and said unto her, If thou knewest the gift of God, and who it is that saith to thee, Give me to drink; thou wouldest have asked of him, and he would have given thee living water.*

11   *The woman saith unto him, Sir, thou hast nothing to draw with, and the well is deep: from whence then hast thou that living water?*

12	*Art thou greater than our father Jacob, which gave us the well, and drank thereof himself, and his children, and his cattle?*
13	*Jesus answered and said unto her, Whosoever drinketh of this water shall thirst again:*
14	*But whosoever drinketh of the water that I shall give him shall never thirst; but the water that I shall give him shall be in him a well of water springing up into everlasting life.*
19	*The woman saith unto him, Sir, I perceive that thou art a prophet.*
23	*But the hour cometh, and now is, when the true worshippers shall worship the Father in spirit and in truth: for the Father seeketh such to worship him.*
24	*God is a Spirit: and they that worship him must worship him in spirit and in truth.*
25	*The woman saith unto him, I know that Messias cometh, which is called Christ: when he is come, he will tell us all things.*
26	*Jesus saith unto her, I that speak unto thee am he.*
28	*The woman then left her waterpot, and went her way into the city, and saith to the men,*
29	*Come, see a man, which told me all things that ever I did: is not this the Christ?*

Christa continues her testimony to David,

"Christ, as the author of all life on earth, has the power to restore human health and life. But Christ's divine power threatened the Senior Jewish Leaders who could not accept and honor Him as the Messiah, the Son of God, as we read in John Chapter 5."

John 5	
18	*Therefore the Jews sought the more to kill him, because he not only had broken the sabbath, but said also that God was his Father, making himself equal with God.*
19	*Then answered Jesus and said unto them, Verily, verily, I say unto you, The Son can do nothing of himself, but what he seeth the Father do: for what things soever he doeth, these also doeth the Son likewise.*
20	*For the Father loveth the Son, and sheweth him all things that himself doeth: and he will shew him greater works than these, that ye may marvel.*
21	*For as the Father raiseth up the dead, and quickeneth them; even so the Son quickeneth whom he will.*
22	*For the Father judgeth no man, but hath committed all judgment unto the Son:*
23	*That all men should honour the Son, even as they honour the Father. He that honoureth not the Son honoureth not the Father which hath sent him.*

Christa and David discuss the role that Christ has with His Father. Christa declares her testimony,

"Christ was edified by His Father, Elohim, and as Elohim can raise the dead, so can His Son restore mortal life and bestow eternal life and

117

exaltation upon those He chooses. Christ informed the Jews that His Father committed the judgment of all mortals to His divine Son as shown in the following verses in John that states that men are resurrected, judged, and assigned a glory by the Son, Jesus Christ.

John 5

24 *Verily, verily, I say unto you, He that heareth my word, and believeth on him that sent me, hath everlasting life, and shall not come into condemnation; but is passed from death unto life.*

25 *Verily, verily, I say unto you, The hour is coming, and now is, when the dead shall hear the voice of the Son of God: and they that hear shall live.*

26 *For as the Father hath life in himself; so hath he given to the Son to have life in himself;*

27 *And hath given him authority to execute judgment also, because he is the Son of man.*

28 *Marvel not at this: for the hour is coming, in the which all that are in the graves shall hear his voice,*

29 *And shall come forth; they that have done good, unto the resurrection of life; and they that have done evil, unto the resurrection of damnation.*

30 *I can of mine own self do nothing: as I hear, I judge: and my judgment is just; because I seek not mine own will, but the will of the Father which hath sent me.*

Christa informs David that Elohim, whom the Jews acknowledge as their God, has given His Son, Jesus Christ, the Messiah sought by the Jews, authority to execute judgment and resurrection of all mankind. Christa then testifies to David that,

"The witness for Christ's divinity is His Father, Elohim who has declared in these Scriptures that eternal life comes through acceptance of the Son, Jesus Christ."

Christa then recalls for David, the state of his being during his NDE and says,

"Christ as the organizer of the Earth has control over all terrestrial and celestial elements and understands all physical laws that govern the behavior of the earth and the universes that comprise all existence. Christ demonstrated that He can exist outside the dimensional constraints that mortals have in time and space. Christ defied earth's gravity by walking above the water on the Sea of Galilee."

Christa reminds David that he was not constrained by gravity and moved about freely through space and time during his NDE. Christa reads to David the following passage from John, Chapter 6 that declares that Christ could exercise control over gravity by walking on the Sea of Galilee.

John 6

17   And entered into a ship, and went over the sea toward Capernaum. And it was now dark, and Jesus was not come to them.

18   And the sea arose by reason of a great wind that blew.

19   So when they had rowed about five and twenty or thirty furlongs, they see Jesus walking on the sea, and drawing nigh unto the ship: and they were afraid.

20   But he saith unto them, It is I; be not afraid.

21   Then they willingly received him into the ship: and immediately the ship was at the land whither they went.

Christa and David discuss that Christ is the omniscient, omnipotent scientist. Christ comprehends the full nature and behavior of all space, time and matter and energy throughout our universe and all other universes that Christ's Father governs. The field of cosmology that David's seeks to explore and understand is completely recognized and comprehended by Christ. Christ can breech the veil or membrane that separates our mortal universe from the higher space and time occupied by His Father to perform the works of His Father, Elohim. Christa then declares that Christ is the true bread of life and those who come to Christ shall never hunger for knowledge and thirst for the truth of all dimensions and time.   Christ's disciples shall be fulfilled as given in John 6.

John 6

27   Labour not for the meat which perisheth, but for that meat which endureth unto everlasting life, which the Son of man shall give unto you: for him hath God the Father sealed.

28   Then said they unto him, What shall we do, that we might work the works of God?

29   Jesus answered and said unto them, This is the work of God, that ye believe on him whom he hath sent.

30   They said therefore unto him, What sign shewest thou then, that we may see, and believe thee? what dost thou work?

> 31 Our fathers did eat manna in the desert; as it is written, He gave them bread from heaven to eat.
>
> 32 Then Jesus said unto them, Verily, verily, I say unto you, Moses gave you not that bread from heaven; but my Father giveth you the true bread from heaven.
>
> 33 For the bread of God is he which cometh down from heaven, and giveth life unto the world.
>
> 34 Then said they unto him, Lord, evermore give us this bread.
>
> 35 And Jesus said unto them, I am the bread of life: he that cometh to me shall never hunger; and he that believeth on me shall never thirst.
>
> 36 But I said unto you, That ye also have seen me, and believe not.
>
> 37 All that the Father giveth me shall come to me; and him that cometh to me I will in no wise cast out.
>
> 38 For I came down from heaven, not to do mine own will, but the will of him that sent me.
>
> 39 And this is the Father's will which hath sent me, that of all which he hath given me I should lose nothing, but should raise it up again at the last day.
>
> 40 And this is the will of him that sent me, that everyone which seeth the Son, and believeth on him, may have everlasting life: and I will raise him up at the last day.

Christa discusses with David that Christ, as the carrier of both mortal genomes from His earthly mother, Mary, and immortal genomes from His Heavenly Father, had within Himself the power to bestow mortals with immortality and to control His life and death. Christa solemnly testifies to David,

"It is Christ's immortal body that contains the immortal genomes and has, as the bread of life, the power to bestow or imbue mortals with immortality as declared in John, Chapter 6."

John 6

> 47 Verily, verily, I say unto you, He that believeth on me hath everlasting life.
>
> 48 I am that bread of life.
>
> 49 Your fathers did eat manna in the wilderness, and are dead.
>
> 50 This is the bread which cometh down from heaven, that a man may eat thereof, and not die.
>
> 51 I am the living bread which came down from heaven: if any man eat of this bread, he shall live forever: and the bread that I will give is my flesh, which I will give for the life of the world.
>
> 53 Then Jesus said unto them, Verily, verily, I say unto you, Except ye eat the flesh of the Son of man, and drink his blood, ye have no life in you.
>
> 54 Whoso eateth my flesh, and drinketh my blood, hath eternal life; and I will raise him up at the last day.

> 55 *For my flesh is meat indeed, and my blood is drink indeed.*
> 56 *He that eateth my flesh, and drinketh my blood, dwelleth in me, and I in him.*
> 57 *As the living Father hath sent me, and I live by the Father: so he that eateth me, even he shall live by me.*
> 58 *This is that bread which came down from heaven: not as your fathers did eat manna, and are dead: he that eateth of this bread shall live forever.*

Christa tells David that the desire to seek a personal testimony of Christ is given to David by God the Father, Elohim. Christa reads that the Jewish Apostle, Peter, who was also a member of the House of David, acknowledged that Christ was the Messiah sought by the Jews and the only begotten Son in the Flesh of His Father. This truth was proclaimed by Peter as is stated in John Chapter 6 and 7.

> **John 6** .
> 65 *And he said, Therefore said I unto you, that no man can come unto me, except it were given unto him of my Father.*
> 66 *From that time many of his disciples went back, and walked no more with him.*
> 67 *Then said Jesus unto the twelve, Will ye also go away?*
> 68 *Then Simon Peter answered him, Lord, to whom shall we go? thou hast the words of eternal life.*
> 69 *And we believe and are sure that thou art that Christ, the Son of the living God.*

Christa and David continue their discussion that the Jews did not believe that Christ was their Messiah. Great turmoil surrounding the Savior and His mission arose as Christ disputed with those Jews who opposed Him, We find this in the following verses in John 7.

> **John 7**
> 14 *Now about the midst of the feast Jesus went up into the temple, and taught.*
> 15 *And the Jews marvelled, saying, How knoweth this man letters, having never learned?*
> 16 *Jesus answered them, and said, My doctrine is not mine, but his that sent me.*
> 17 *If any man will do his will, he shall know of the doctrine, whether it be of God, or whether I speak of myself.*
> 18 *He that speaketh of himself seeketh his own glory: but he that seeketh his glory that sent him, the same is true, and no unrighteousness is in him.*

> 19 Did not Moses give you the law, and yet none of you keepeth the law? Why go ye about to kill me?
> 20 The people answered and said, Thou hast a devil: who goeth about to kill thee?
> 21 Jesus answered and said unto them, I have done one work, and ye all marvel.
> 22 Moses therefore gave unto you circumcision; (not because it is of Moses, but of the fathers;) and ye on the sabbath day circumcise a man.
> 23 If a man on the sabbath day receive circumcision, that the law of Moses should not be broken; are ye angry at me, because I have made a man every whit whole on the sabbath day?
> 24 Judge not according to the appearance, but judge righteous judgment.
> 25 Then said some of them of Jerusalem, Is not this he, whom they seek to kill?
> 26 But, lo, he speaketh boldly, and they say nothing unto him. Do the rulers know indeed that this is the very Christ?
> 27 Howbeit we know this man whence he is: but when Christ cometh, no man knoweth whence he is.
> 31 And many of the people believed on him, and said, When Christ cometh, will he do more miracles than these which this man hath done?

Christa testifies to David that the final and absolute witness for Christ's divinity is His Father, Elohim who has declared in the Scriptures that Christ is His only Begotten Son in the Flesh and eternal life for man comes only through acceptance of His Son, Jesus Christ.

> **John 7**
> 37 In the last day, that great day of the feast, Jesus stood and cried, saying, If any man thirst, let him come unto me, and drink.
> 38 He that believeth on me, as the scripture hath said, out of his belly shall flow rivers of living water.
> 40 Many of the people therefore, when they heard this saying, said, Of a truth this is the Prophet.
> 41 Others said, This is the Christ. But some said, Shall Christ come out of Galilee?
> 42 Hath not the scripture said, That Christ cometh of the seed of David, and out of the town of Bethlehem, where David was?
> 43 So there was a division among the people because of him.

Christa tells David that Christ is the Light of the World. David responds to Christa that light is the common term for all frequencies of electromagnetic radiation carried by photons, which are the bosons associated with electromagnetic radiation. Significantly, light is the principal means by which physics examines the Universe in which we live. Indeed, our eyes that receive visible light, are the major sources of man's everyday knowledge of his surroundings. Therefore, since Christ is the Light of the world, He can dispel our ignorance of darkness.

David tells Christa that cosmology now realizes that 95% of the energy and mass necessary to account for the observed behavior in our visible universe is not observable. This deficit is composed of dark matter at ~27% and dark energy at ~68%, neither of which is physically observable in our physical universe. These dark entities may exist in higher dimensions that are not observable in our present spacetime universe. Christa responds that Christ, however, has knowledge of and control over this dark matter and energy since all dimensions and times are open to Christ and His Father, Elohim. Christa and David then discuss that since Christ is the light of the world, He is the true source by which mankind will comprehend heaven and earth, and it is through His Gospel that we can become heirs to our Heavenly Father's Kingdom and experience eternal life. Christa reads from John 8 the scripture confirming that Christ is the Light of the world and the true Messiah expected by the Jews.

John 8

12 Then spake Jesus again unto them, saying, I am the light of the world: he that followeth me shall not walk in darkness, but shall have the light of life.

13 The Pharisees therefore said unto him, Thou bearest record of thyself; thy record is not true.

14 Jesus answered and said unto them, Though I bear record of myself, yet my record is true: for I know whence I came, and whither I go; but ye cannot tell whence I come, and whither I go.

15 Ye judge after the flesh; I judge no man.

16 And yet if I judge, my judgment is true: for I am not alone, but I and the Father that sent me.

17 It is also written in your law, that the testimony of two men is true.

18 I am one that bear witness of myself, and the Father that sent me beareth witness of me.

19 Then said they unto him, Where is thy Father? Jesus answered, Ye neither know me, nor my Father: if ye had known me, ye should have known my Father also.

---

24 *I said therefore unto you, that ye shall die in your sins: for if ye believe not that I am he, ye shall die in your sins.*
28 *Then said Jesus unto them, When ye have lifted up the Son of Man, then shall ye know that I am he, and that I do nothing of myself; but as my Father hath taught me, I speak these things.*
29 *And he that sent me is with me: the Father hath not left me alone; for I do always those things that please him.*
30 *As he spake these words, many believed on him.*
31 *Then said Jesus to those Jews which believed on him, If ye continue in my word, then are ye my disciples indeed;*
32 *And ye shall know the truth, and the truth shall make you free.*

Christa tells David that when he accepts the words and Gospel of Christ, then David will become a Disciple of Christ and will know the truth that Christ is the Son of God, the Savior of the World. This truth will make David free to receive Eternal Life. Christa then tells David that David's early progenitor, Abraham, acknowledged Christ as the Messiah by declaring,

"Another witness for Christ's divinity was Abraham, the Patriarch for the House of Israel. Abraham rejoiced to see the coming of Christ, who is the Messiah as we read in the following verses in John, Chapter 8."

John 8 .
33 *They answered him, We be Abraham's seed, and were never in bondage to any man: how sayest thou, Ye shall be made free?*
34 *Jesus answered them, Verily, verily, I say unto you, Whosoever committeth sin is the servant of sin.*
35 *And the servant abideth not in the house forever: but the Son abideth ever.*
36 *If the Son therefore shall make you free, ye shall be free indeed.*
37 *I know that ye are Abraham's seed; but ye seek to kill me, because my word hath no place in you.*
38   *I speak that which I have seen with my Father: and ye do that which ye have seen with your father.*
39 *They answered and said unto him, Abraham is our father. Jesus saith unto them, If ye were Abraham's children, ye would do the works of Abraham.*
40 *But now ye seek to kill me, a man that hath told you the truth, which I have heard of God: this did not Abraham.*
41 *Ye do the deeds of your father. Then said they to him, We be not born of fornication; we have one Father, even God.*
42   *Jesus said unto them, If God were your Father, ye would love me: for I proceeded forth and came from God; neither came I of myself, but he sent me.*

---

51 Verily, verily, I say unto you, If a man keep my saying, he shall never see death.

52 Then said the Jews unto him, Now we know that thou hast a devil. Abraham is dead, and the prophets; and thou sayest, If a man keep my saying, he shall never taste of death.

53 Art thou greater than our father Abraham, which is dead? and the prophets are dead: whom makest thou thyself?

54 Jesus answered, If I honour myself, my honour is nothing: it is my Father that honoureth me; of whom ye say, that he is your God:

55 Yet ye have not known him; but I know him: and if I should say, I know him not, I shall be a liar like unto you: but I know him, and keep his saying.

56 Your father Abraham rejoiced to see my day: and he saw it, and was glad.

57 Then said the Jews unto him, Thou art not yet fifty years old, and hast thou seen Abraham?

58 Jesus said unto them, Verily, verily, I say unto you, Before Abraham was, I am.

59 Then took they up stones to cast at him: but Jesus hid himself, and went out of the temple, going through the midst of them, and so passed by.

Christa informs David of the many miracles wrought by Christ that demonstrated His divinity and power, including restoration of the sight of one born blind as declared in John 9. Only Christ had the power to give vision to one who never had eyesight.

John 9

1 And as Jesus passed by, he saw a man which was blind from his birth.

5 As long as I am in the world, I am the light of the world.

6 When he had thus spoken, he spat on the ground, and made clay of the spittle, and he anointed the eyes of the blind man with the clay,

7 And said unto him, Go, wash in the pool of Siloam, (which is by interpretation, Sent.) He went his way therefore, and washed, and came seeing.

10 Therefore said they unto him, How were thine eyes opened?

11 He answered and said, A man that is called Jesus made clay, and anointed mine eyes, and said unto me, Go to the pool of Siloam, and wash: and I went and washed, and I received sight.

14 And it was the sabbath day when Jesus made the clay, and opened his eyes.

15 Then again the Pharisees also asked him how he had received his sight. He said unto them, He put clay upon mine eyes, and I washed, and do see.

16 Therefore said some of the Pharisees, This man is not of God, because he keepeth not the sabbath day. Others said, How can a man that is a sinner do such miracles? And there was a division among them.

17 They say unto the blind man again, What sayest thou of him, that he hath opened thine eyes? He said, He is a prophet.

---

18 *But the Jews did not believe concerning him, that he had been blind, and received his sight, until they called the parents of him that had received his sight.*

19 *And they asked them, saying, Is this your son, who ye say was born blind? how then doth he now see?*

20 *His parents answered them and said, We know that this is our son, and that he was born blind:*

24 *Then again called they the man that was blind, and said unto him, Give God the praise: we know that this man is a sinner.*

25 *He answered and said, Whether he be a sinner or no, I know not: one thing I know, that, whereas I was blind, now I see.*

28 *Then they reviled him, and said, Thou art his disciple; but we are Moses' disciples.*

29 *We know that God spake unto Moses: as for this fellow, we know not from whence he is*

30 *The man answered and said unto them, Why herein is a marvelous thing, that ye know not from whence he is, and yet he hath opened mine eyes.*

31 *Now we know that God heareth not sinners: but if any man be a worshipper of God, and doeth his will, him he heareth.*

32 *Since the world began was it not heard that any man opened the eyes of one that was born blind.*

33 *If this man were not of God, he could do nothing.*

34 *They answered and said unto him, Thou wast altogether born in sins, and dost thou teach us? And they cast him out.*

35 *Jesus heard that they had cast him out; and when he had found him, he said unto him, Dost thou believe on the Son of God?*

36 *He answered and said, Who is he, Lord, that I might believe on him?*

37 *And Jesus said unto him, Thou hast both seen him, and it is he that talketh with thee.*

38 *And he said, Lord, I believe. And he worshipped him.*

Christa tells David of her belief that the spittle with which Christ anointed the blind man contained some of Christ's unique genomes that regenerated the eyes of the blind man just as the natural eyes are formed from the human zygote.

They then discuss that Christ's sheep, His followers, will recognize Christ as their shepherd and will join with Him. Furthermore, Christ declared that there are other sheep who dwelled not in Jerusalem but in the America's and elsewhere as told in the Book of Mormon.

John 10

14 I am the good shepherd, and know my *sheep,* and am known of mine.

15 As the Father knoweth me, even so know I the Father: and I lay down my life for the sheep.

16 And other sheep I have, which are not of this fold: them also I must bring, and they shall hear my voice; and there shall be one fold, *and* one shepherd.

17 Therefore doth my Father love me, because I lay down my life, that I might take it again.

18 No man taketh it from me, but I lay it down of myself. I have power to lay it down, and I have power to take it again. This commandment have I received of my Father.

19 There was a division therefore again among the Jews for these sayings.

20 And many of them said, He hath a devil, and is mad; why hear ye him?

21 Others said, These are not the words of him that hath a devil. Can a devil open the eyes of the blind?

Christa tells David that Mormons believe that other sheep referenced by Christ included those inhabitants of the Americas that Christ visited after His crucifixion and resurrection as documented in the Book of Mormon.

Despite the many miracles performed by Christ, some Jews would not accept him as the Messiah or the Christ. Jesus repeatedly declared His Messiahship, but the Jews did not believe Him as recorded in John 10.

John 10

24 Then came the Jews round about him, and said unto him, How long dost thou make us to doubt? If thou be the Christ, tell us plainly.

25 Jesus answered them, I told you, and ye believed not: the works that I do in my Father's name, they bear witness of me.

26 But ye believe not, because ye are not of my sheep, as I said unto you.

27 My sheep hear my voice, and I know them, and they follow me:

28 And I give unto them eternal life; and they shall never perish, neither shall any *man* pluck them out of my hand.

29 My Father, which gave *them* me, is greater than all; and no *man* is able to pluck *them* out of my Father's hand.

30 I and *my* Father are one.

31 Then the Jews took up stones again to stone him.

32 Jesus answered them, Many good works have I shewed you from my Father; for which of those works do ye stone me?

33 The Jews answered him, saying, For a good work we stone thee not; but for blasphemy; and because that thou, being a man, makest thyself God.

34 Jesus answered them, Is it not written in your law, I said, Ye are gods?

> 35 If he called them gods, unto whom the word of God came, and the scripture cannot be broken;
> 36 Say ye of him, whom the Father hath sanctified, and sent into the world, Thou blasphemest; because I said, I am the Son of God?
> 37 If I do not the works of my Father, believe me not.
> 38 But if I do, though ye believe not me, believe the works: that ye may know, and believe, that the Father *is* in me, and I in him.

Christa tells David in the 11th Chapter of John, Christ raised Lazarus, a devout Jew from the grave. Lazarus had been in the grave for four days demonstrating Christ's power over mortal death. This seminal event greatly disturbed certain members of the Pharisees and Sadducees, because this event caused many of the common Jews to recognize that Jesus was the Christ, the Messiah, that Jews have sought throughout their history.

> **John 11**
> *1 Now a certain man was sick, named Lazarus, of Bethany, the town of Mary and her sister Martha.*
> *3 Therefore his sisters sent unto him, saying, Lord, behold, he whom thou lovest is sick.*
> *4 When Jesus heard that, he said, This sickness is not unto death, but for the glory of God, that the Son of God might be glorified thereby.*
> *7 Then after that saith he to his disciples, Let us go into Judæa again.*
> *11 These things said he: and after that he saith unto them, Our friend Lazarus sleepeth; but I go, that I may awake him out of sleep.*
> *12 Then said his disciples, Lord, if he sleep, he shall do well.*
> *13 Howbeit Jesus spake of his death: but they thought that he had spoken of taking of rest in sleep.*
> *14 Then said Jesus unto them plainly, Lazarus is dead.*
> *15 And I am glad for your sakes that I was not there, to the intent ye may believe; nevertheless let us go unto him.*
> *17 Then when Jesus came, he found that he had lain in the grave four days already.*
> *21 Then said Martha unto Jesus, Lord, if thou hadst been here, my brother had not died.*
> *22 But I know, that even now, whatsoever thou wilt ask of God, God will give it thee.*
> *23 Jesus saith unto her, Thy brother shall rise again.*
> *24 Martha saith unto him, I know that he shall rise again in the resurrection at the last day.*
> *25 Jesus said unto her, I am the resurrection, and the life: he that believeth in me, though he were dead, yet shall he live:*
> *26 And whosoever liveth and believeth in me shall never die. Believest thou this?*
> *27 She saith unto him, Yea, Lord: I believe that thou art the Christ, the Son of God, which should come into the world.*

28 *And when she had so said, she went her way, and called Mary her sister secretly, saying, The Master is come, and calleth for thee.*

29 *As soon as she heard that, she arose quickly, and came unto him.*

32 *Then when Mary was come where Jesus was, and saw him, she fell down at his feet, saying unto him, Lord, if thou hadst been here, my brother had not died.*

33 *When Jesus therefore saw her weeping, and the Jews also weeping which came with her, he groaned in the spirit, and was troubled,*

34 *And said, Where have ye laid him? They said unto him, Lord, come and see.*

35 *Jesus wept.*

36 *Then said the Jews, Behold how he loved him!*

37 *And some of them said, Could not this man, which opened the eyes of the blind, have caused that even this man should not have died?*

38 *Jesus therefore again groaning in himself cometh to the grave. It was a cave, and a stone lay upon it.*

39 *Jesus said, Take ye away the stone. Martha, the sister of him that was dead, saith unto him, Lord, by this time he stinketh: for he hath been dead four days.*

40 *Jesus saith unto her, Said I not unto thee, that, if thou wouldest believe, thou shouldest see the glory of God?*

41 *Then they took away the stone from the place where the dead was laid. And Jesus lifted up his eyes, and said, Father, I thank thee that thou hast heard me.*

42 *And I knew that thou hearest me always: but because of the people which stand by I said it, that they may believe that thou hast sent me.*

43 *And when he thus had spoken, he cried with a loud voice, Lazarus, come forth.*

44 *And he that was dead came forth, bound hand and foot with grave clothes: and his face was bound about with a napkin. Jesus saith unto them, Loose him, and let him go.*

45 *Then many of the Jews which came to Mary, and had seen the things which Jesus did, believed on him.*

Christa informs David that Christ's authority and miracles threatened the Jewish leaders, so they sought to take his life. These leaders feared they would lose their position with the Roman rulers in Jerusalem. We are told this in John 11, when the Jewish High Priest, Caiaphas, sought the death of Jesus

John 11

47 *Then gathered the chief priests and the Pharisees a council, and said, What do we? for this man doeth many miracles.*

48 *If we let him thus alone, all men will believe on him: and the Romans shall come and take away both our place and nation.*

49 *And one of them, named Caiaphas, being the high priest that same year, said unto them, Ye know nothing at all,*

> 50   Nor consider that it is expedient for us, that one man should die for the people, and that the whole nation perish not.
> 53   Then from that day forth they took counsel together for to put him to death.

Christa tells David that many of the Jewish people believed upon Christ and hailed Him as the King of Israel and the Messiah upon His triumphal entry into Jerusalem. However, the Jewish leaders were threatened by the growing acceptance of Christ, also called the Son of Man, and the public raising of Lazarus from the grave.

John 12

12   On the next day much people that were come to the feast, when they heard that Jesus was coming to Jerusalem,

13   Took branches of palm trees, and went forth to meet him, and cried, Hosanna: Blessed is the King of Israel that cometh in the name of the Lord.

14   And Jesus, when he had found a young ass, sat thereon; as it is written,

15   Fear not, daughter of Sion: behold, thy King cometh, sitting on an ass's colt.

17   The people therefore that was with him when he called Lazarus out of his grave, and raised him from the dead, bare record.

18   For this cause the people also met him, for that they heard that he had done this miracle.

19   The Pharisees therefore said among themselves, Perceive ye how ye prevail nothing? behold, the world is gone after him.

21   The same came therefore to Philip, which was of Bethsaida of Galilee, and desired him, saying, Sir, we would see Jesus.

22   Philip cometh and telleth Andrew: and again Andrew and Philip tell Jesus.

23   And Jesus answered them, saying, The hour is come, that the Son of Man should be glorified.

24   Verily, verily, I say unto you, Except a corn of wheat fall into the ground and die, it abideth alone: but if it die, it bringeth forth much fruit.

26   If any man serve me, let him follow me; and where I am, there shall also my servant be: if any man serve me, him will my Father honour.

27   Now is my soul troubled; and what shall I say? Father, save me from this hour: but for this cause came I unto this hour.

28   Father, glorify thy name. Then came there a voice from heaven, saying, I have both glorified it, and will glorify it again.

29   The people therefore, that stood by, and heard it, said that it thundered: others said, An angel spake to him.

30   Jesus answered and said, This voice came not because of me, but for your sakes.

31   Now is the judgment of this world: now shall the prince of this world be cast out.

32   And I, if I be lifted up from the earth, will draw all men unto me.

33   This he said, signifying what death he should die.

34 The people answered him, We have heard out of the law that Christ abideth forever: and how sayest thou, The Son of Man must be lifted up? who is this Son of Man?

35 Then Jesus said unto them, Yet a little while is the light with you. Walk while ye have the light, lest darkness come upon you: for he that walketh in darkness knoweth not whither he goeth.

36 While ye have light, believe in the light, that ye may be the children of light. These things spake Jesus, and departed, and did hide himself from them.

37 But though he had done so many miracles before them, yet they believed not on him:

44 Jesus cried and said, He that believeth on me, believeth not on me, but on him that sent me.

45 And he that seeth me seeth him that sent me.

46 I am come a light into the world, that whosoever believeth on me should not abide in darkness.

47 And if any man hear my words, and believe not, I judge him not: for I came not to judge the world, but to save the world.

48 He that rejecteth me, and receiveth not my words, hath one that judgeth him: the word that I have spoken, the same shall judge him in the last day.

49 For I have not spoken of myself; but the Father which sent me, he gave me a commandment, what I should say, and what I should speak.

50 And I know that his commandment is life everlasting: whatsoever I speak therefore, even as the Father said unto me, so I speak.

Christ knew that He would be slain for His words and deeds as told in John 13. So the Savior prepared the Apostles for His death and resurrection by washing the feet of His Apostles as a sign of His service and love for the Apostles.

John 13

1 Now before the feast of the passover, when Jesus knew that his hour was come that he should depart out of this world unto the Father, having loved his own which were in the world, he loved them unto the end.

2 And supper being ended, the devil having now put into the heart of Judas Iscariot, Simon's son, to betray him;

3 Jesus knowing that the Father had given all things into his hands, and that he was come from God, and went to God;

4 He riseth from supper, and laid aside his garments; and took a towel, and girded himself.

5 After that he poureth water into a bason, and began to wash the disciples' feet, and to wipe them with the towel wherewith he was girded.

6 Then cometh he to Simon Peter: and Peter saith unto him, Lord, dost thou wash my feet?

7 Jesus answered and said unto him, What I do thou knowest not now; but thou shalt know hereafter.

8   *Peter saith unto him, Thou shalt never wash my feet. Jesus answered him, If I wash thee not, thou hast no part with me.*

9   *Simon Peter saith unto him, Lord, not my feet only, but also my hands and my head.*

10   *Jesus saith to him, He that is washed needeth not save to wash his feet, but is clean every whit: and ye are clean, but not all.*

12   *So after he had washed their feet, and had taken his garments, and was set down again, he said unto them, Know ye what I have done to you?*

13   *Ye call me Master and Lord: and ye say well; for so I am.*

14   *If I then, your Lord and Master, have washed your feet; ye also ought to wash one another's feet.*

15   *For I have given you an example, that ye should do as I have done to you.*

16   *Verily, verily, I say unto you, The servant is not greater than his lord; neither he that is sent greater than he that sent him.*

17   *If ye know these things, happy are ye if ye do them.*

18   *I speak not of you all: I know whom I have chosen: but that the scripture may be fulfilled, He that eateth bread with me hath lifted up his heel against me.*

19   *Now I tell you before it come, that, when it is come to pass, ye may believe that I am he.*

20   *Verily, verily, I say unto you, He that receiveth whomsoever I send receiveth me; and he that receiveth me receiveth him that sent me.*

21   *When Jesus had thus said, he was troubled in spirit, and testified, and said, Verily, verily, I say unto you, that one of you shall betray me.*

22   *Then the disciples looked one on another, doubting of whom he spake.*

23   *Now there was leaning on Jesus' bosom one of his disciples, whom Jesus loved.*

24   *Simon Peter therefore beckoned to him, that he should ask who it should be of whom he spake.*

25   *He then lying on Jesus' breast saith unto him, Lord, who is it?*

26   *Jesus answered, He it is, to whom I shall give a sop, when I have dipped it. And when he had dipped the sop, he gave it to Judas Iscariot, the son of Simon.*

27   *And after the sop Satan entered into him. Then said Jesus unto him, That thou doest, do quickly.*

28   *Now no man at the table knew for what intent he spake this unto him.*

29   *For some of them thought, because Judas had the bag, that Jesus had said unto him, Buy those things that we have need of against the feast; or, that he should give something to the poor.*

30   *He then having received the sop went immediately out: and it was night.*

31   *Therefore, when he was gone out, Jesus said, Now is the Son of man glorified, and God is glorified in him.*

32   *If God be glorified in him, God shall also glorify him in himself, and shall straightway glorify him.*

33   *Little children, yet a little while I am with you. Ye shall seek me: and as I said unto the Jews, Whither I go, ye cannot come; so now I say to you.*

34   *A new commandment I give unto you, That ye love one another; as I have loved you, that ye also love one another.*

35   *By this shall all men know that ye are my disciples, if ye have love one to another.*

> 36 Simon Peter said unto him, Lord, whither goest thou? Jesus answered him, Whither I go, thou canst not follow me now; but thou shalt follow me afterwards.
>
> 37 Peter said unto him, Lord, why cannot I follow thee now? I will lay down my life for thy sake.
>
> 38 Jesus answered him, Wilt thou lay down thy life for my sake? Verily, verily, I say unto thee, The cock shall not crow, till thou hast denied me thrice.

Christa tells David that since Christ was the Only Begotten of the Father in the Flesh, Christ is in the very image of His Father as a mortal son can be in the image of His mortal father as told in John 14. Thus, Christ and His Father, have the image of men.

John 14

1 Let not your heart be troubled: ye believe in God, believe also in me.

2 In my Father's house are many mansions: if it were not so, I would have told you. I go to prepare a place for you.

3 And if I go and prepare a place for you, I will come again, and receive you unto myself; that where I am, there ye may be also.

4 And whither I go ye know, and the way ye know.

5 Thomas saith unto him, Lord, we know not whither thou goest; and how can we know the way?

6 Jesus saith unto him, I am the way, the truth, and the life: no man cometh unto the Father, but by me.

7 If ye had known me, ye should have known my Father also: and from henceforth ye know him, and have seen him.

8 Philip saith unto him, Lord, shew us the Father, and it sufficeth us.

9 Jesus saith unto him, Have I been so long time with you, and yet hast thou not known me, Philip? he that hath seen me hath seen the Father; and how sayest thou then, Shew us the Father?

10 Believest thou not that I am in the Father, and the Father in me? the words that I speak unto you I speak not of myself: but the Father that dwelleth in me, he doeth the works.

11 Believe me that I am in the Father, and the Father in me: or else believe me for the very works' sake.

12 Verily, verily, I say unto you, He that believeth on me, the works that I do shall he do also; and greater works than these shall he do; because I go unto my Father.

13 And whatsoever ye shall ask in my name, that will I do, that the Father may be glorified in the Son.

14 If ye shall ask any thing in my name, I will do it.

15 If ye love me, keep my commandments.

16 And I will pray the Father, and he shall give you another Comforter, that he may abide with you forever;

> 17 Even the Spirit of truth; whom the world cannot receive, because it seeth him not, neither knoweth him: but ye know him; for he dwelleth with you, and shall be in you.
>
> 18 I will not leave you comfortless: I will come to you.
>
> 19 Yet a little while, and the world seeth me no more; but ye see me: because I live, ye shall live also.
>
> 20 At that day ye shall know that I am in my Father, and ye in me, and I in you.
>
> 21 He that hath my commandments, and keepeth them, he it is that loveth me: and he that loveth me shall be loved of my Father, and I will love him, and will manifest myself to him.

Christa discusses with David that Christ told the Apostles that His Father would send the Holy Ghost as a Comforter and Substitute for Christ when He departed the earth as we read in John 14 and 15.

> **John 14**
>
> 26 But the Comforter, which is the Holy Ghost, whom the Father will send in my name, he shall teach you all things, and bring all things to your remembrance, whatsoever I have said unto you.
>
> 27 Peace I leave with you, my peace I give unto you: not as the world giveth, give I unto you. Let not your heart be troubled, neither let it be afraid.
>
> 28 Ye have heard how I said unto you, I go away, and come again unto you. If ye loved me, ye would rejoice, because I said, I go unto the Father: for my Father is greater than I.
>
> **John 15**
>
> 26 But when the Comforter is come, whom I will send unto you from the Father, even the Spirit of truth, which proceedeth from the Father, he shall testify of me:
>
> 27 And ye also shall bear witness, because ye have been with me from the beginning.

In John 16 Christa states that Jesus told His followers of the mission of the Holy Ghost. Also, Christ reaffirms His death and resurrection and again declares He is the Son of God.

> **John 16**
>
> 7 Nevertheless I tell you the truth; It is expedient for you that I go away: for if I go not away, the Comforter will not come unto you; but if I depart, I will send him unto you.
>
> 8 And when he is come, he will reprove the world of sin, and of righteousness, and of judgment:
>
> 9 Of sin, because they believe not on me;
>
> 10 Of righteousness, because I go to my Father, and ye see me no more;

11   *Of judgment, because the prince of this world is judged.*

12   *I have yet many things to say unto you, but ye cannot bear them now.*

13   *Howbeit when he, the Spirit of truth, is come, he will guide you into all truth: for he shall not speak of himself; but whatsoever he shall hear, that shall he speak: and he will shew you things to come.*

14   *He shall glorify me: for he shall receive of mine, and shall shew it unto you.*

15   *All things that the Father hath are mine: therefore said I, that he shall take of mine, and shall shew it unto you.*

16   *A little while, and ye shall not see me: and again, a little while, and ye shall see me, because I go to the Father.*

17   *Then said some of his disciples among themselves, What is this that he saith unto us, A little while, and ye shall not see me: and again, a little while, and ye shall see me: and, Because I go to the Father?*

18   *They said therefore, What is this that he saith, A little while? we cannot tell what he saith.*

19   *Now Jesus knew that they were desirous to ask him, and said unto them, Do ye enquire among yourselves of that I said, A little while, and ye shall not see me: and again, a little while, and ye shall see me?*

20   *Verily, verily, I say unto you, That ye shall weep and lament, but the world shall rejoice: and ye shall be sorrowful, but your sorrow shall be turned into joy.*

21   *A woman when she is in travail hath sorrow, because her hour is come: but as soon as she is delivered of the child, she remembereth no more the anguish, for joy that a man is born into the world.*

22   *And ye now therefore have sorrow: but I will see you again, and your heart shall rejoice, and your joy no man taketh from you.*

23   *And in that day ye shall ask me nothing. Verily, verily, I say unto you, Whatsoever ye shall ask the Father in my name, he will give it you.*

24   *Hitherto have ye asked nothing in my name: ask, and ye shall receive, that your joy may be full.*

25   *These things have I spoken unto you in proverbs: but the time cometh, when I shall no more speak unto you in proverbs, but I shall shew you plainly of the Father.*

26   *At that day ye shall ask in my name: and I say not unto you, that I will pray the Father for you:*

27   *For the Father himself loveth you, because ye have loved me, and have believed that I came out from God.*

28   *I came forth from the Father, and am come into the world: again, I leave the world, and go to the Father.*

29   *His disciples said unto him, Lo, now speakest thou plainly, and speakest no proverb.*

30   *Now are we sure that thou knowest all things, and needest not that any man should ask thee: by this we believe that thou camest forth from God.*

31   *Jesus answered them, Do ye now believe?*

32   *Behold, the hour cometh, yea, is now come, that ye shall be scattered, every man to his own, and shall leave me alone: and yet I am not alone, because the Father is with me.*

33   *These things I have spoken unto you, that in me ye might have peace. In the world ye shall have tribulation: but be of good cheer; I have overcome the world.*

Christa then solemnly reads to David the great intercessory prayer. Christ consummated the matchless Atonement while in the Garden of Gethsemane. He relinquished His will to the Father. He declared that eternal life is gained by recognizing and honoring the Father, Elohim, and becoming a Disciple of Jesus Christ whom the Father sent. Christa then testifies to David, that she earnestly wishes David to become a Disciple of Christ.

John 17

1   *These words spake Jesus, and lifted up his eyes to heaven, and said, Father, the hour is come; glorify thy Son, that thy Son also may glorify thee:*

2   *As thou hast given him power over all flesh, that he should give eternal life to as many as thou hast given him.*

3   *And this is life eternal, that they might know thee the only true God, and Jesus Christ, whom thou hast sent.*

4   *I have glorified thee on the earth: I have finished the work which thou gavest me to do.*

5   *And now, O Father, glorify thou me with thine own self with the glory which I had with thee before the world was.*

6   *I have manifested thy name unto the men which thou gavest me out of the world: thine they were, and thou gavest them me; and they have kept thy word.*

7   *Now they have known that all things whatsoever thou hast given me are of thee.*

8   *For I have given unto them the words which thou gavest me; and they have received them, and have known surely that I came out from thee, and they have believed that thou didst send me.*

9   *I pray for them: I pray not for the world, but for them which thou hast given me; for they are thine.*

10 *And all mine are thine, and thine are mine; and I am glorified in them.*

11 *And now I am no more in the world, but these are in the world, and I come to thee. Holy Father, keep through thine own name those whom thou hast given me, that they may be one, as we are.*

12 *While I was with them in the world, I kept them in thy name: those that thou gavest me I have kept, and none of them is lost, but the son of perdition; that the scripture might be fulfilled.*

13 *And now come I to thee; and these things I speak in the world, that they might have my joy fulfilled in themselves.*

14 *I have given them thy word; and the world hath hated them, because they are not of the world, even as I am not of the world.*

15 *I pray not that thou shouldest take them out of the world, but that thou shouldest keep them from the evil.*

16 *They are not of the world, even as I am not of the world.*

17 *Sanctify them through thy truth: thy word is truth.*

---

18 As thou hast sent me into the world, even so have I also sent them into the world.
19 And for their sakes I sanctify myself, that they also might be sanctified through the truth.
20 Neither pray I for these alone, but for them also which shall believe on me through their word;
21 That they all may be one; as thou, Father, art in me, and I in thee, that they also may be one in us: that the world may believe that thou hast sent me.
22 And the glory which thou gavest me I have given them; that they may be one, even as we are one:
23 I in them, and thou in me, that they may be made perfect in one; and that the world may know that thou hast sent me, and hast loved them, as thou hast loved me.
24 Father, I will that they also, whom thou hast given me, be with me where I am; that they may behold my glory, which thou hast given me: for thou lovedst me before the foundation of the world.
25 O righteous Father, the world hath not known thee: but I have known thee, and these have known that thou hast sent me.
26 And I have declared unto them thy name, and will declare it: that the love wherewith thou hast loved me may be in them, and I in them.

Christa informs David that after Christ's atoning sacrifice in the Garden of Gethsemane, He was betrayed, arrested and arraigned before Pilate, the Roman leader, to suffer crucifixion. Although the Sanhedrin sought Christ's death, only the Roman Procreator could render such a final judgment.

John 18
1 When Jesus had spoken these words, he went forth with his disciples over the brook Cedron, where was a garden, into the which he entered, and his disciples.
2 And Judas also, which betrayed him, knew the place: for Jesus ofttimes resorted thither with his disciples.
3 Judas then, having received a band of men and officers from the chief priests and Pharisees, cometh thither with lanterns and torches and weapons.
4 Jesus therefore, knowing all things that should come upon him, went forth, and said unto them, Whom seek ye?
5 They answered him, Jesus of Nazareth. Jesus saith unto them, I am he. And Judas also, which betrayed him, stood with them.
6 As soon then as he had said unto them, I am he, they went backward, and fell to the ground.
7 Then asked he them again, Whom seek ye? And they said, Jesus of Nazareth.
8 Jesus answered, I have told you that I am he: if therefore ye seek me, let these go their way:

9   *That the saying might be fulfilled, which he spake, Of them which thou gavest me have I lost none.*

10 *Then Simon Peter having a sword drew it, and smote the high priest's servant, and cut off his right ear. The servant's name was Malchus.*

11 *Then said Jesus unto Peter, Put up thy sword into the sheath: the cup which my Father hath given me, shall I not drink it?*

12 *Then the band and the captain and officers of the Jews took Jesus, and bound him,*

13 *And led him away to Annas first; for he was father in law to Caiaphas, which was the high priest that same year.*

14 *Now Caiaphas was he, which gave counsel to the Jews, that it was expedient that one man should die for the people.*

19 *The high priest then asked Jesus of his disciples, and of his doctrine.*

20 *Jesus answered him, I spake openly to the world; I ever taught in the synagogue, and in the temple, whither the Jews always resort; and in secret have I said nothing.*

21 *Why askest thou me? ask them which heard me, what I have said unto them: behold, they know what I said.*

28 *Then led they Jesus from Caiaphas unto the hall of judgment: and it was early; and they themselves went not into the judgment hall, lest they should be defiled; but that they might eat the passover.*

29 *Pilate then went out unto them, and said, What accusation bring ye against this man?*

30 *They answered and said unto him, If he were not a malefactor, we would not have delivered him up unto thee.*

31 *Then said Pilate unto them, Take ye him, and judge him according to your law. The Jews therefore said unto him, It is not lawful for us to put any man to death:*

32 *That the saying of Jesus might be fulfilled, which he spake, signifying what death he should die.*

33 *Then Pilate entered into the judgment hall again, and called Jesus, and said unto him, Art thou the King of the Jews?*

34 *Jesus answered him, Sayest thou this thing of thyself, or did others tell it thee of me?*

35 *Pilate answered, Am I a Jew? Thine own nation and the chief priests have delivered thee unto me: what hast thou done?*

36 *Jesus answered, My kingdom is not of this world: if my kingdom were of this world, then would my servants fight, that I should not be delivered to the Jews: but now is my kingdom not from hence.*

37 *Pilate therefore said unto him, Art thou a king then? Jesus answered, Thou sayest that I am a king. To this end was I born, and for this cause came I into the world, that I should bear witness unto the truth. Every one that is of the truth heareth my voice.*

38 *Pilate saith unto him, What is truth? And when he had said this, he went out again unto the Jews, and saith unto them, I find in him no fault at all.*

Christa somberly tells David that Jesus Christ, the very Son of God and the Messiah, was scourged, crucified and then buried in the tomb of

Joseph of Arimathea. We learn of this savage treatment of Christ as detailed in John 19.

John 19

1   Then Pilate therefore took Jesus, and scourged him.

2   And the soldiers platted a crown of thorns, and put it on his head, and they put on him a purple robe,

3   And said, Hail, King of the Jews! and they smote him with their hands.

4   Pilate therefore went forth again, and saith unto them, Behold, I bring him forth to you, that ye may know that I find no fault in him.

5   Then came Jesus forth, wearing the crown of thorns, and the purple robe. And Pilate saith unto them, Behold the man!

6   When the chief priests therefore and officers saw him, they cried out, saying, Crucify him, crucify him. Pilate saith unto them, Take ye him, and crucify him: for I find no fault in him.

7   The Jews answered him, We have a law, and by our law he ought to die, because he made himself the Son of God.

8   When Pilate therefore heard that saying, he was the more afraid;

9   And went again into the judgment hall, and saith unto Jesus, Whence art thou? But Jesus gave him no answer.

10   Then saith Pilate unto him, Speakest thou not unto me? knowest thou not that I have power to crucify thee, and have power to release thee?

11   Jesus answered, Thou couldest have no power at all against me, except it were given thee from above: therefore he that delivered me unto thee hath the greater sin.

12   And from thenceforth Pilate sought to release him: but the Jews cried out, saying, If thou let this man go, thou art not Cæsar's friend: whosoever maketh himself a king speaketh against Cæsar.

13   When Pilate therefore heard that saying, he brought Jesus forth, and sat down in the judgment seat in a place that is called the Pavement, but in the Hebrew, Gabbatha.

14   And it was the preparation of the passover, and about the sixth hour: and he saith unto the Jews, Behold your King!

15   But they cried out, Away with him, away with him, crucify him. Pilate saith unto them, Shall I crucify your King? The chief priests answered, We have no king but Cæsar.

16   Then delivered he him therefore unto them to be crucified. And they took Jesus, and led him away.

17   And he bearing his cross went forth into a place called the place of a skull, which is called in the Hebrew Golgotha:

18   Where they crucified him, and two other with him, on either side one, and Jesus in the midst.

19   And Pilate wrote a title, and put it on the cross. And the writing was, JESUS OF NAZARETH THE KING OF THE JEWS.

20   This title then read many of the Jews: for the place where Jesus was crucified was nigh to the city: and it was written in Hebrew, and Greek, and Latin.

---

21 Then said the chief priests of the Jews to Pilate, Write not, The King of the Jews; but that he said, I am King of the Jews.
22 Pilate answered, What I have written I have written.
28 After this, Jesus knowing that all things were now accomplished, that the scripture might be fulfilled, saith, I thirst.
29 Now there was set a vessel full of vinegar: and they filled a spunge with vinegar, and put it upon hyssop, and put it to his mouth.
30 When Jesus therefore had received the vinegar, he said, It is finished: and he bowed his head, and gave up the ghost.
31 The Jews therefore, because it was the preparation, that the bodies should not remain upon the cross on the sabbath day, (for that sabbath day was an high day,) besought Pilate that their legs might be broken, and that they might be taken away.
32 Then came the soldiers, and brake the legs of the first, and of the other which was crucified with him.
33 But when they came to Jesus, and saw that he was dead already, they brake not his legs:
34 But one of the soldiers with a spear pierced his side, and forthwith came there out blood and water.
38 And after this Joseph of Arimathæa, being a disciple of Jesus, but secretly for fear of the Jews, besought Pilate that he might take away the body of Jesus: and Pilate gave him leave. He came therefore, and took the body of Jesus.
39 And there came also Nicodemus, which at the first came to Jesus by night, and brought a mixture of myrrh and aloes, about an hundred pound weight.
40 Then took they the body of Jesus, and wound it in linen clothes with the spices, as the manner of the Jews is to bury.
41 Now in the place where he was crucified there was a garden; and in the garden a new sepulchre, wherein was never man yet laid.
42 There laid they Jesus therefore because of the Jews' preparation day; for the sepulchre was nigh at hand.

Christa then solemnly and tearfully testifies to David that since Christ was the true Messiah, He was resurrected, and His resurrection was manifest for all mankind. This unique and glorious event was witnessed by Mary Magdalene and later by the Disciples as told in John 20.

John 20
1 The first day of the week cometh Mary Magdalene early, when it was yet dark, unto the sepulchre, and seeth the stone taken away from the sepulchre.
2 Then she runneth, and cometh to Simon Peter, and to the other disciple, whom Jesus loved, and saith unto them, They have taken away the Lord out of the sepulchre, and we know not where they have laid him.
3 Peter therefore went forth, and that other disciple, and came to the sepulchre.

4   So they ran both together: and the other disciple did outrun Peter, and came first to the sepulchre.

5   And he stooping down, and looking in, saw the linen clothes lying; yet went he not in.

6   Then cometh Simon Peter following him, and went into the sepulchre, and seeth the linen clothes lie,

7   And the napkin, that was about his head, not lying with the linen clothes, but wrapped together in a place by itself.

8   Then went in also that other disciple, which came first to the sepulchre, and he saw, and believed.

9   For as yet they knew not the scripture, that he must rise again from the dead.

10   Then the disciples went away again unto their own home.

11   But Mary stood without at the sepulchre weeping: and as she wept, she stooped down, and looked into the sepulchre,

12   And seeth two angels in white sitting, the one at the head, and the other at the feet, where the body of Jesus had lain.

13   And they say unto her, Woman, why weepest thou? She saith unto them, Because they have taken away my Lord, and I know not where they have laid him.

14   And when she had thus said, she turned herself back, and saw Jesus standing, and knew not that it was Jesus.

15   Jesus saith unto her, Woman, why weepest thou? whom seekest thou? She, supposing him to be the gardener, saith unto him, Sir, if thou have borne him hence, tell me where thou hast laid him, and I will take him away.

16   Jesus saith unto her, Mary. She turned herself, and saith unto him, Rabboni; which is to say, Master.

17   Jesus saith unto her, Touch me not; for I am not yet ascended to my Father: but go to my brethren, and say unto them, I ascend unto my Father, and your Father; and to my God, and your God.

18   Mary Magdalene came and told the disciples that she had seen the Lord, and that he had spoken these things unto her.

19   Then the same day at evening, being the first day of the week, when the doors were shut where the disciples were assembled for fear of the Jews, came Jesus and stood in the midst, and saith unto them, Peace be unto you.

20   And when he had so said, he shewed unto them his hands and his side. Then were the disciples glad, when they saw the Lord.

21   Then said Jesus to them again, Peace be unto you: as my Father hath sent me, even so send I you.

22   And when he had said this, he breathed on them, and saith unto them, Receive ye the Holy Ghost:

23   Whose soever sins ye remit, they are remitted unto them; and whose soever sins ye retain, they are retained.

24   But Thomas, one of the twelve, called Didymus, was not with them when Jesus came.

25   The other disciples therefore said unto him, We have seen the Lord. But he said unto them, Except I shall see in his hands the print of the nails, and put my finger into the print of the nails, and thrust my hand into his side, I will not believe.

26	*And after eight days again his disciples were within, and Thomas with them: then came Jesus, the doors being shut, and stood in the midst, and said, Peace be unto you.*
27	*Then saith he to Thomas, Reach hither thy finger, and behold my hands; and reach hither thy hand, and thrust it into my side: and be not faithless, but believing.*
28	*And Thomas answered and said unto him, My Lord and my God.*
29	*Jesus saith unto him, Thomas, because thou hast seen me, thou hast believed: blessed are they that have not seen, and yet have believed.*
30	*And many other signs truly did Jesus in the presence of his disciples, which are not written in this book:*
31	*But these are written, that ye might believe that Jesus is the Christ, the Son of God; and that believing ye might have life through his name.*

Christa and David discuss that the resurrected Christ appeared again to His disciples and charged them to declare his Gospel during the forty days that Christ was with His apostles before His ascension to His Father. Christa soberly testifies to David that she is deeply committed to declaring this Gospel to David. David will be resurrected, and he can gain eternal life if he accepts and loves Christ and seeks to keep His commandments and follow Him as declared in John 21.

John 21	
1.	*After these things Jesus shewed himself again to the disciples at the sea of Tiberias; and on this wise shewed he himself.*
2.	*There were together Simon Peter, and Thomas called Didymus, and Nathanael of Cana in Galilee, and the sons of Zebedee, and two other of his disciples.*
3.	*Simon Peter saith unto them, I go a fishing. They say unto him, We also go with thee. They went forth, and entered into a ship immediately; and that night they caught nothing.*
4.	*But when the morning was now come, Jesus stood on the shore: but the disciples knew not that it was Jesus.*
5.	*Then Jesus saith unto them, Children, have ye any meat? They answered him, No.*
6.	*And he said unto them, Cast the net on the right side of the ship, and ye shall find. They cast therefore, and now they were not able to draw it for the multitude of fishes.*
7.	*Therefore that disciple whom Jesus loved saith unto Peter, It is the Lord. Now when Simon Peter heard that it was the Lord, he girt his fisher's coat unto him, (for he was naked,) and did cast himself into the sea.*
9.	*As soon then as they were come to land, they saw a fire of coals there, and fish laid thereon, and bread.*
10.	*Jesus saith unto them, Bring of the fish which ye have now caught.*
11.	*Simon Peter went up, and drew the net to land full of great fishes, an hundred and fifty and three: and for all there were so many, yet was not the net broken.*

12. *Jesus saith unto them, Come and dine. And none of the disciples durst ask him, Who art thou? knowing that it was the Lord.*
13. *Jesus then cometh, and taketh bread, and giveth them, and fish likewise.*
14. *This is now the third time that Jesus shewed himself to his disciples, after that he was risen from the dead.*
15. *So when they had dined, Jesus saith to Simon Peter, Simon, son of Jonas, lovest thou me more than these? He saith unto him, Yea, Lord; thou knowest that I love thee. He saith unto him, Feed my lambs.*
16. *He saith to him again the second time, Simon, son of Jonas, lovest thou me? He saith unto him, Yea, Lord; thou knowest that I love thee. He saith unto him, Feed my sheep*
17. *He saith unto him the third time, Simon, son of Jonas, lovest thou me? Peter was grieved because he said unto him the third time, Lovest thou me? And he said unto him, Lord, thou knowest all things; thou knowest that I love thee. Jesus saith unto him, Feed my sheep.*

Christa testifies to David:

"Christ closed his mortal ministry and charged His Apostles to feed his sheep. By this command, Christ directed His Apostles to promulgate the Good News of the Gospel to all who would hear and accept Christ as their Savior. Those who accept His Gospel and are baptized become His Disciples and will dwell with Him and His Father, Elohim, in the Kingdom of Heaven."

Then with tears in her eyes, Christa tenderly declares,

"David, I want you to become a Disciple of Christ with me, so we can mutually live the Gospel of Christ and receive Eternal life through His Atonement. David, I want a Disciple of Christ as my eternal companion. Will you receive Christ and become His disciple?"

David realizes that he must carefully ponder these solemn words spoken by Christa and her invitation to become a disciple of Christ. David's growing dependence upon and love for Christa pose a challenge for him to satisfy Christa's tender demand and his own Jewish traditions and background.

David's conclusions regarding M-Theory (2Dec)

During their extensive interchange regarding the Gospel of Christ, David expands his investigations and conclusions regarding M-theory and the Gospel that he shares on a continuous basis with Christa. David tells Christa of his expanded model for the universe. There are at least two universes, and probably many more as envision by "cosmic inflation," that enshroud one another, but are separated by a membrane or barrier. Cosmologists term these universes as parallel universes since they can interact with each other, but only across this membrane. The one universe, our familiar four-dimensional, spacetime domain, is the Universe in which we live and move as earthly mortals. A parallel universe that adjoins our universe has different properties and physical laws for its constituents including more spatial dimensions and a bi-directional dimension for time. We are familiar with our universe since it contains our planet earth, the sun and the all the other planets in the solar system as well as the Milky Way Galaxy and hundreds of billions of other galaxies that reside in our observable universe. Our solar system is a minor constituent in the Milky Way Galaxy that has over a hundred billion stars with orbiting planets and forms a spiral disk with a diameter of about 100,000 light years. Although these very large numbers are difficult for humans to comprehend, we now know that there are hundreds of billion other comparable galaxies similar to our Milky Way Galaxy within our observable universe.

The complete physical and chemical composition of our observable Universe is now beginning to unfold before David. Our universe is a vast, rapidly expanding spherical spacetime assembly with a radius of at least 13.8 billion light years and hundreds of billions of galaxies and hundreds of billions of stars within each galaxy. The dimension we call time in our galaxy is a consequence of and controlled by the Second Law of Thermodynamics. This Law controls the unilateral forward flow of time through the continual increase in entropy for our Universe. Randomness in the gross structure and the order of matter and energy in our universe is constantly increasing. Entropy is generally associated with large ensembles of matter, but it is evident even in the world of Quantum Mechanics. However, in isolation, the

Second Law poses a dilemma for our Universe. The Big Bang event that occurred about 13.8 billion years ago started at very low; near zero level of Entropy for our Universe. The quandary is what and how did this very low-level entropy state originate in time before the Big Bang that birthed our Universe if entropy can only increase within any closed system. The Second Law appears to control and dominate our Universe immediately following the Big Bang since all observable natural processes in our Universe exhibit a continual increase of Entropy with time. The critical question and dilemma is how then was the cosmological egg for the Big Bang that spawned our Universe brought into being? Was some other parallel universe the source that served to provide the egg that birthed our own four spacetime universe at a very low value of Entropy that has now been increasing ever since the Big Bang?

A credible answer is that this low entropy state was introduced from outside our Universe by another universe…a parallel universe composed of dark matter that surrounds our own Universe, but is separated by a Membrane or "brane". Our Universe is not an isolated, closed system, but is directly coupled to this parallel universe. This assumption would appear to simply delay the argument of how the parallel universe began if that parallel universe is also subject to the Second Law of Thermodynamics or Entropy. However, what is also possible is to assert that in the parallel universe of dark matter, the Second Law of Thermodynamics is not operable, as we know it. Time and space, the familiar three-space and one-time dimensions and many of the laws of physics and chemistry are not the same and operate differently in this parallel universe that surrounds us in a larger parallel universe composed most of dark matter.

It is also possible that extra dimensions exist in a parallel universe in which most, perhaps many or all of the observations and theories of space, time, and matter that we claim in our universe are different in this parallel universe. In M-Theory, there are ten dimensions in space and one in time and the extra dimensions may not be hidden in that parallel universe, as they appear to be in our four-spacetime universe. In addition, time in this parallel universe may not be the same as we reckon time in our universe. Time for us is unilateral, and flows only in one direction…only forward so we age, and decay, and die. All matter and energy in our universe degrades and is dispersed with increase in entropy. However, our enshrouding, parallel universe may not be constrained by the forward flow of time. Indeed, the forward singular flow of time as we understand the passage of time may exist only for us in our present universe. It is possible that the laws of physics and the nature and behavior of matter and energy as we observe them are very different in this parallel universe.

David's theories of physical models (2Dec)

David tells Christa that to spawn this original cosmic egg for the birth of our universe, the parallel universe may have intersected our universe in spacetime at some location on the brane separating the two universes and this occurred at the beginning of the big bang event 13.8 billion years earlier in our time. Time may be completely open and not transitory or forward directional in the parallel universe but arbitrary and without real meaning or time measurement as we know it. Time in the parallel universe may be similar to open space in our own universe. In the parallel universe, the dimension of time may be like the dimension for space. Movement in any direction forward or backward in time or remain at a fixed time as we perceive time in our universe may exist. When David was in his near-death experience, he was possibly a temporary visitor in that parallel universe. Time as perceived by him then was an open dimension and he could move back and forth through time while he was in that parallel universe. He could move and select any location or any time for observing events, that had occurred, were occurring, or would occur in our universe.

Furthermore, human thoughts and desires as well as actions and language were discernible and evident to him as he witnessed the accident that claimed the life of his sister. These properties of complete and accurate observation and comprehension are those we ascribe to God and His omniscience and omnipotence. The Christian God is held to be without limits and constraints in either time or space. God can access and control events at any location and any time in our universe. He dwells in the parallel universe in which His access and control of our spacetime Universe is completely open and fully accessible. The challenge for David is that although he personally witnessed this evidence in his near-death experience, how could he establish the existence of this parallel universe or Kingdom of Heaven to unbelieving critics, especially his scientific colleagues? Must humans die before they can recognize what David has witnessed and knows exist? David asserts that the concept and reality of a parallel universe is essential for explaining the 'Big Bang' event in our universe. As the Bible states, in the beginning God created the heavens and the earth. From God's perspective in His parallel universe, He could observe such events and proclaim the assessment "…and God saw that it was good." These extraordinary claims and David's NDE could merge Quantum Mechanics and Einstein Mechanics and more importantly, explain and confirm the existence and the nature of God and His role in our universe.

Christa and David continue discussions when Christa returns to apartment (22 Oct)

Christa is delighted with David's work and its compatibility and support for the Gospel of Jesus Christ. Often in the evening when Christa has returned to the apartment, she and David will continue their pursuit of the scriptures. Christa often reads a scripture that she wants to share with David. David enjoys these close and intimate times and enjoys hearing the Christa's eloquent words that sooth him and build the rapport and understanding they share with one another. They mutually discuss the meaning and intent of Gospel words and impact upon their lives.

David progresses to wheel chair (22 Oct)

David's physical condition improves through extensive medical treatment and intense physical therapy. He is able to sit in a wheel chair during the day and access his Apple Laptop Computer on a worktable where he can display his papers and deliberate on his mathematical explorations. During David's solitary and focused research work sessions, Christa attends to her nursing duties and then sits quietly near David while she reads Scriptures and responds to their mutual mail and financial matters. David enjoys her close, but silent presence and recognizes Christa's respect for his focused concentration and support for his research.

Christa adroitly adjusts to her role as a gentle observer and tacit supporter while David pursues his research. During these quiet, tender and intimate settings, Christa ponders silently in her mind that their future life together as a married couple would be what she experiences now…a peaceful, fulfilling companionship for Christa with a scientific genius at work.

David's and classical violin music (22 Oct)

David requires diversions from his intense research sessions and religious discussions especially in the evenings when Christa is absent from his Hospital Suite or otherwise unavailable. David recognizes that it is during those diversionary times when he is alone that his subconscious mind advances new and novel approaches to his research and the scriptures discussed with Christa. These inward times often prove to be very effective in providing new insights, understanding and spiritual and scientific development.

David has expressed to Christa, his deep enjoyment of background, classical violin music that he particularly enjoys in the evening as he ponders his research and Gospel discussions during the day with Christa. To address David's wishes, Christa obtains an AP file with classical violin music for David to play on his I-Pod during her absence. Although David enjoys most of the great violin concerti by the masters such as Mozart, Brahms, Tchaikovsky, Mendelssohn, and Paganini, the unique piece he venerates most is the famous Violin Concerto in D by Ludwig Von Beethoven. The Rondo (Allegro) movement from this Classic Concerto, David particularly reveres. Furthermore, although he also acknowledges many fine performing artists of this unique classical violin work, his favorite contemporary performer is the renowned Jewish violinist, Yitzhak Perlman. There are others concert violinists who are also exceptional such as Isaac Stern, the young Joshua Bell and the incomparable Jascha Heifetz who is now deceased. David informed Christa that Heifetz traveled to Israel to support a concert when David was a child. David attended the concert with his father and Rebecca. A contemporary associate of Heifetz played the incomparable Rondo movement and David has revered that particular violin concerto ever since that memorable performance. When David listens to Beethoven's Violin Concerto now, fond memories are aroused of his deceased father whom he idolized. When David listens to the strains of the Violin Concerto in D, he is a young boy with his beloved father feasting together on this violin classic.

David also realizes that it is during these solitary evenings in his Hospital Suite while listening to his music and allowing his mind to drift that he develops new insights and inspiration regarding his work in M-Theory. Most important, it is during these solitary moments that his increasing understanding and acceptance of the Gospel of Jesus Christ is imperceptibly deepening in his mind. During these times his thoughts are unconstrained and true understanding and creativity as well as spiritual awakening occurs.

Christa records in her journal the intense discussions between David and herself regarding the Gospel, particularly as recorded in John and also David's instruction and their discussions regarding David's research in M-Theory and its relevance to the Gospel.

149

Christa's weekly Personal Journal for 25-31 Oct 2010

I am delighted with the progress that David and I have made in reading and discussing the scriptures together, particularly those in the Gospel of John in the New Testament. Also, David is educating me in his research and the extensions it has with the Gospel and David's development in M-Theory. I am confident that David now understands and will accept the Gospel. Our extensive scriptural studies and spiritual discussions have given David a growing testimony of Christ as the Messiah, the one promised to rescue the Jewish people and deliver all mankind from death.

David's pursuit of the Gospel and the association of his NDE and his research into M-Theory have all combined to forge David's testimony and acceptance of Christ as the Son of God. As a result of these wonderful, spiritual developments guided by the Holy Ghost and answers to my fervent prayers, David will become a Disciple of Christ with me.

In closing my journal entry, I am resolved and confident that David is now the man in my life who will become my husband and eternal companion. Bless me to that end, Father if that is Thy will also.

CHAPTER 8 (1 Nov-7 Nov)

Bomb Incident at Hospital (1Nov)

Upon Christa's arrival in David's Hospital Suite on this Monday morning, he immediately notifies her,

"Christa, someone with an explosive device just passed by my Hospital Suite this morning. The device must be a small, crude contrivance composed of ammonium nitrate and a detonator. Ammonium compounds have very characteristic aromatic signatures. A mass of a fraction of a kilogram of ammonium nitrate can cause serious damage if successfully detonated. Although limited in destructive range, this device still poses sufficient damage to kill and injure persons within the vicinity of the explosion.

Christa, upon hearing this warning from David, immediately informs hospital security of the bomb threat. All patients in suites in the area of David's Hospital Suite on the fourth floor and on the third and fifth floor are quickly evacuated. The Hospital Security Network is rapidly deployed and bomb squad members begin an immediate sweep of the Hospital suites near David's Hospital Suite on the fourth floor of Massachusetts General Hospital. However, the surgical floor is on the sixth floor and cannot be immediately evacuated because of ongoing surgeries.

Two orderlies, accompanied by Christa, transport David in his bed and his portable monitoring equipment along the hospital corridor of the fourth floor away from David's Hospital Suite towards the elevators at the far end of the corridor.

As they move David in his hospital bed along the corridor, David calls to Christa.

"Christa, the explosive device is in the linen storage room that we just passed."

Christa tells the attendants to move David to the far end of the corridor while she informs Security of the possible location of the bomb.

Two members of the bomb squad quickly enter the linen storage room with surveillance dogs trained to detect explosives. The bomb squad soon discovers a crude explosive device composed of ammonium nitrate and a timer wrapped in hospital towels and wedged in the upper shelf of the bed linen closet. The squad carefully removes the device and defuses the bomb. No other bomb materials are found in the storage room.

Bomb diffused (1Nov)

After the bomb is diffused and further, exhaustive search of the fourth floor is conducted, the incident is safely resolved, and patients are returned to their suites. As the emergency alert is ended, the Administrator of Massachusetts General Hospital informs the staff and patients over the hospital PA system that Christa Olsen, the fourth-floor nurse for David Steinmann, discovered the bomb and the swift response of the bomb team terminated this emergency without harm to any of the patients or staff at the Hospital.

Upon hearing the announcement, Christa wants to inform the Administrator who actually identified and found the device, but David restrains her.

"Christa, It was your rapid and effective notification to the proper personnel that resolved the incident. Furthermore, I cannot disclose my sensory ability to others. It might compromise my defense research with the Israeli and US Government."

An investigation is initiated to determine the source and identity of the terrorist. Subsequent investigation shows that the bomb was placed in the linen closet adjacent to the hospital suite of a hemorrhaging female patient recently admitted from an illegal abortion clinic. An investigation is instigated to determine the identity of the assailant.

David's second surgery (3Nov)

As David progresses in his surgical procedures and physical therapy, he is soon able to walk short distances with a cane, generally with

Christa immediately nearby to ensure his stability. He and Christa slowly walk the corridors of Massachusetts General Hospital to augment the physical exercises associated with the demanding physical therapy prescribed. With Christa's perseverance and care, these serial surgeries and physical therapy are restoring David's strength and ambulatory agility.

NDE impact upon David's religious beliefs (5 Nov)

David confides to Christa that his NDE has provided him with a rational basis for belief in God, and in the reality of another higher dimensional universe surrounding our universe. David believes that this experience and further investigation will support his pursuits in M-Theory. David is convinced that his NDE experience, although surreal, was accurate and pivotal in support of his work in M-Theory. He actually witnessed during his NDE access and unrestrained freedom of movement in both space and time. He was able to move through space and time at will and without the physical constraints that we encounter in our conscious physical universe.

David informs Christa of his thoughts and aspirations regarding the universe and our position within it,

"I have been pondering especially if time as we know and measure time, may have different properties in other universes. The properties of matter in our universe and the laws of physics that we have developed here on Earth may be very different in a parallel universe. Physicists have pondered that if the electrical properties and masses of fundamental particles such as the electron and proton differ, even by a small amount, the consequences would be dramatic. If such differences exist, then different and remarkable properties and conditions may exist in any parallel universe."

Christa and David discuss at length the significance of time as measured and observed by God. David's NDE has shown for him that time may be open in the Kingdom of Heaven. Time there is like a spatial dimension and can be accessed forward, backward or in the present. Events in time are completely open before the observer.

Christa reads from the scriptures that she has accessed that address time. She reads the following verses and they discuss that time is a different measure for God.

Alma 40.8 ...all is as one day with God, and time only is measured unto man.
D&C 84.100 ...And Satan is bound and time is no longer.
D&C 130.4 & 5 ...Is not the reckoning of God's time, angel's time, prophet's time, and man's time, according to the planet on which they reside? I answer, Yes.
Abraham 3.4 ...Kolob was after the manner of the Lord, according to its times and seasons in the revolutions thereof; that one revolution was a day unto the Lord, and his manner of reckoning, it being one thousand years according to the time appointed unto that whereon thou standest. This is the reckoning of the Lord's time.

Hospital Candy striper (5 Nov)

David is frequently visited in his Hospital Suite by a Candy Stripper at the Hospital whose name is Maria Martinez. Christa informs David that Maria is a fifteen-year-old female Hispanic orphan who lives with her Grandmother in a low-cost tenement district outside Boston. Maria is a serious student who visits the hospital after school several days a week and spends most of her visitation time with the elderly patients who receive few visitors to their rooms. Living with her elderly Grandmother, Maria is sensitive to the needs of the elderly. Since Maria is conversant in English as well as Spanish, the Hispanic Hospital residents find her particularly helpful since she can communicate freely with English and Spanish personnel.

Maria picks small bouquets of sunflowers and other wild flowers she finds in vacant lots near her dwelling and brings these welcome floral offerings to many of the patients she visits. Her cheerful disposition, diminutive presence and childlike innocence make her a delightful visitor at the Hospital. David and Christa have formed a strong affection and concern for this young girl who they recognize is very poor and needy, but generous with her time and love of others. Christa pays Maria generously for the flowers she brings to the patients with David's encouragement. Christa artfully tells Maria that her floral offerings save Christa from shopping for flowers for the patients on the fourth floor where David is located. The flowers bring cheerful smiles to the patients who anticipate Maria's visits and her kindness.

David and Christa observe that Maria always wears the same simple gingham, one-piece cotton dress for her hospital visits. Since she comes directly after school, they realize she must wear these same clothes to school each day. David and Christa discuss Maria's limited economic circumstances, and David asks Christa to take Maria shopping on the following Saturday and buy her a new wardrobe of clothes for school. Christa is delighted with David's suggestion and agrees to spend Saturday with Maria.

Christa takes Maria shopping (6 Nov) inspect tour

To Maria's delight, Christa invites Maria to visit the City of Boston on Saturday. Christa and Maria meet at a mutual METRO station on the following Saturday, and Christa takes Maria shopping to buy new clothes and have lunch in the city. Maria and Christa explore a variety of department stores and shops. Christa assesses Maria's choices although Maria is reluctant to express favorites since she realizes the expense of attractive clothes for young women. With Christa's insistence, they return to their final clothing store of choice and Christa helps Maria to select a full wardrobe including undergarments, two delightful teenage skirts and matching blouses, new shoes and matching stockings. Christa also insists that Maria choose an attractive Sunday outfit for attendance at Maria's Catholic Mass. After their exhausting but delightful shopping spree, they have lunch at a sidewalk café where Maria experiences for the first time in her young life the crowds and bustle of downtown Boston on a cool winter day.

During their lunch, Maria confides to Christa that she has visited Boston on infrequent school tours. These supervised visits were restricted to libraries and museums under the supervision of her teacher. Maria is openly excited and deeply grateful to Christa for this special time, attention, and most of all for the new clothes she has acquired in Boston. Maria now believes she will make new friends at school because of her improved appearance among her classmates.

Christa and Maria then travel together on the METRO to return Maria home to her Grandmother. Christa visits with the Grandmother and learns more of the family circumstances. Maria's mother and father died in Cuba, and Maria was sent by remaining relatives to America for her Grandmother to rear and support.

Christa's Journal entry regarding occupancy in David's apartment (7 Nov Sunday)

Christa and David continue their established regular evening scriptural discussions using their individual BlackBerrys after Christa returns to the apartment from the hospital. Christa is delighted with this routine since it is essential evidence that David is gaining a testimony of the Gospel and recognition that Christ is central in both Christa's life and David's future with Christa.

Over the weekend, Christa meditates over the events of the past week. She ponders the success of her efforts to become David's wife and the reasons for her growing love for David. His concern for a poor, Hispanic orphan, Maria, and the joy and satisfaction that David and she realize with our support and assistance for this child demonstrate to Christa that David has strong Christian love for others that is essential for a Disciple of Christ. Although David is a world leader in science and deeply engaged in his research, he demonstrates compassion and support for an orphan girl and her condition in life. Besides a great mind and intellect, David exhibits Christian charity and concern for others. Christa's love for David is fully evident in her mind and heart. Christa is certain David will become her eternal companion after she brings him to the Gospel of Jesus Christ.

Before retiring on Sunday evening and after kneeling in prayer, Christa opens her Journal and writes the following addressing the important events of the past week. Christa believes that David is gaining a testimony of divinity and mission of Jesus Christ.

Christa's weekly Personal Journal for 1-7 Nov 2010

David has an incredible sense of smell that can be embarrassing for me. But, he discovered a bomb at the hospital using his unique sensatory perception. Security located and disabled the bomb because David could sense the presence of ammonium nitrate.

On Saturday, I spent the day with Maria and took her shopping in Boston for new clothes for school and church. She is the young candy stripper on the fourth floor, Maria Martinez. She is an orphan who lives with her aged grandmother. David gave me money to outfit this needy young child and shower her with attention, kindness and some simple material gifts. Maria was ecstatic, and her innocent face beamed as we leisurely shopped in downtown Boston and purchased several clothing outfits and new shoes. Later, I took her home and we visited briefly with her grandmother. It is such a wonderful feeling of Christian service to bless this child's life and witness her joy over acquisition of simple apparel to dress her properly.

I am fascinated with David's many unique attributes and skills, his exceptional mind, his pursuit of a testimony in the Gospel and, of course, his obvious attention for me. I am flattered that he finds me so attractive. David has no previous romantic experience or close female friends, but I will correct that. I will become the singular, most important woman in his life as his companion ... his eternal companion.

Heavenly Father, I now believe that David has a firm testimony of the divinity and mission of Thy Son. Please continue to bless me with success in my efforts to bring him to Thee. David will be a worthy disciple of Christ.

CHAPTER 9 (8 Nov-14 Nov)

Maria visits David and Christa at Hospital (8 Nov Monday)

The following Monday when Maria comes to David's Hospital Suite after school, she is obviously excited to display to David and Christa her new school clothes. She is delighted with her light blue-frocked blouse and matching pleated skirt. These clothes and her shiny black shoes and pink stockings set Maria apart now as a lovely young woman with attire to match. David and Christa compliment Maria on how attractive she is with her new school wardrobe. The joy on her face is evident as she parades before her appreciative audience. Maria confesses that she was so excited to display her new wardrobe that she forgot to gather flowers for David's Hospital Suite, but she promises to bring flowers on her next visit. As Maria departs, David's Hospital Suite, she gives a warm embrace to Christa and David and again thanks them for her new clothes.

While David and Christa discuss Maria and her service for the patients at the Hospital, David candidly informs Christa that Maria is pregnant. Her close presence during her embrace of David evidenced the scent of her active gestation.

Christa is very disturbed with this disclosure, but she has no doubt of David's sensatory assessment. Later, in a private and confidential setting, Christa gently confronts Maria regarding David's disclosure. After a somber, tear filled confession, Maria admits that she has been intimate with a boy in her tenement. She has worried that she might be pregnant since she missed her last menstrual period. The boy who fathered the child wants Maria to have an abortion and tells Maria he knows of a person in the tenement who will perform an abortion for only $100. Christa is greatly disturbed with her words and counsels Maria to reject this serious and perilous offer to terminate her pregnancy. Such a medical procedure is dangerous and must be performed at a competent physician's office under sterile conditions and with proper facilities and medical support. These illegal abortion settings could threaten Maria's life as well as terminate the life of her baby. Christa tells Maria that to abort her fetus is also a grave moral sin in the eyes of God and is a serious violation of her Christian faith as a Catholic. Christa then offers the services of Massachusetts General Hospital to find a family that wants to adopt a baby. Christa counsels Maria that there are many married couples who cannot have children, and they would welcome, love, and provide for her child. Also, the adopting family would fully cover all of her prenatal expenses, help her with counseling, and even secure a teacher to continue her education during the last few months of her pregnancy. If Maria approves, Christa will make the necessary arrangements through the hospital to provide medical care for Maria and adoption for her baby. Finally, Christa will accompany Maria to inform her Grandmother of her pregnancy. After sustained counsel with tender compassion, Maria embraces Christa and with tears, Maria agrees for Christa to arrange for the adoption of the baby and Christa's visit with Maria's Grandmother.

Christa informs David of the results of her meeting and counsel with Maria and the choice she has made for adoption of the baby. David is very grateful that this serious episode in Maria's young life will be satisfactorily resolved. David praises Christa for her compassion and love for Maria, an unfortunate victim and vulnerable child.

Christa and David pray together (12 Nov Friday)

During one of their regular evening exchanges using their BlackBerrys with David in his Hospital Suite and Christa in David's apartment, Christa asks David solemnly,

"David, will you satisfy an important need I have in our developing relationship?"

David responds quickly with humor declaring,

"Certainly Christa, would you wish me to come and tuck you in for the evening?"

Christa recognizing David's levity, counters his petition with her usual, erudite eloquence,

"You realize my dear Professor that would be very difficult considering the time and spatial constraints that confine us in our present four dimensional, inertial, spacetime universe. I do not anticipate that you will experience another NDE that will allow you to travel instantaneously to my location via a parallel universe. As a conscious mortal, sequestered in our present universe, you are constrained to travel at less than light speed to favor me."

Christa, then returns to her sincere request, and provocatively entreats David,

"See, David, I am becoming conversant with your strange, ethereal world of physics and I hope my soliloquy confirms my efforts to investigate your realm in science. Seriously David, I have pondered how to approach you with my request; but I will just declare my proposition boldly and directly and hope you will be compliant and conciliatory to my sincere request."

David responds thoughtfully,

"Christa, there are few requests that I could or would deny you; so speak on eloquently, Nurse Olsen."

Hesitantly Christa utters in subdued words,

"David…I want us…I want us to pray together using our BlackBerrys when I retire for the evening in the apartment. It will be a Christian prayer and I will be the voice for our oblation. I will pray to 'Our Heavenly Father' and endorse our supplication in the name of Jesus Christ. We will mutually petition our Heavenly Father as we seek to engage Him in our maturing relationship with each other and your testimony of the Gospel. Will you participate with me in this very important act of faith? It would mean a great deal to me. As you know, I have been praying alone for your full recovery and return to Harvard since I first became your nurse. I have also humbly petitioned our Heavenly Father that you will gain a compelling testimony of the Savior, Jesus Christ, and enter the waters of Baptism. I believe that my personal prayers are being answered in our behalf; but I now need and would cherish your participation in my prayers...in our mutual prayers together."

A long pause transpires over the BlackBerry as Christa anxiously awaits David's response. Christa realizes the significance of her request and the impact upon David and potential conflict with his Jewish heritage. After several moments of silence, David responds to Christa's anxious words saying softly,

"Christa, if this will please you and magnify our relationship, then I agree. However, realize that prayer, particularly a Christian prayer is unfamiliar to me and I may prove to be an inept participant, at least at first. But I cannot refuse the person in my life that administers to my health, controls my personal affairs, supports my research, arouses my religious conversion to Christianity and, of greatest importance, stirs my passions and evokes my love."

Christa is overjoyed at David's response and is without words to express her gratitude for his approval. Finally, David responds after a sustained pause waiting for Christa's response via his BlackBerry,

"Hopefully, you won't ask me to kneel."

Christa quickly responds then with transparent satisfaction and enlarged confidence in her petition.

"No David, I will not ask you to kneel. However, I do wish we could kneel...together...in prayer. That would also be compelling evidence of your full recovery. Oh, thank you dear David, your approval is the answer to my fervent prayers in our behalf."

Thus continues David's slow and gentle conversion by Christa to the Gospel of Jesus Christ. Christa skillfully and humbly plans and slowly executes the conversion of David ben-Steinmann to her Christian, Mormon faith. Significantly, Christa establishes regular daily prayer in their evening exchanges using their individual, linked BlackBerrys. Christa initially leads their prayer in the evening with her kneeling in David's apartment and David in his bed at the Hospital. Later, they are praying regularly in the evening with Christa as the voice before they retire. Finally, they begin to alternate their oral prayers together using their BlackBerrys for communication. David's oblations are very simple and mimic Christa's words. But, slowly, Christa realizes that David is gaining confidence in and a testimony of the Savior within his own soul.

This sacred event in their lives greatly gratifies Christa, and she believes that the Holy Ghost will now enlarge David's understanding and testimony of the Savior and God's universe and His plans for David. David is gradually engaging in solemn prayer, although he finds this new experience unfamiliar and challenging. But he loves Christa and is prepared to satisfy her earnest requests and needs. Furthermore, unknown to David, Christa's private petitions to the Lord for the Spirit of the Holy Ghost to descend upon David are gradually coming to past. The deepening love that David has for Christa and the workings of the Spirit are bringing David to a solemn testimony of the Savior and a realization of the truth of Christ's Gospel and acknowledgement of His Atonement in David's behalf.

David's resumes research with students (9 Nov Tuesday)

David wants to resume his graduate colloquium with his graduate students. David confides with Christa that the unwritten truth about pioneering developments in science is the fact that it is the dedicated graduate students who support and perform the major labor in the development in advanced scientific research. These students receive modest financial support and critical stimulation from their major professors, but it is these dedicated students who perform the hard, laborious analysis and spend the long hours of trial and error in seeking answers to the research crusades of their research professors. David has three brilliant and very diligent and resourceful graduate and post-doctoral students who have been assisting him with his research in M-Theory for the last several years. Their commitment and extraordinary skill and labor are essential for the difficult and arduous mathematical modeling and analysis that David supervises.

The mathematical basis of M-Theory is so abstruse and extraordinary that David must often research and develop novel mathematic bases and logic to disclose and support the concepts required to address M-Theory. These mathematical models are very abstract and often not well developed. For example, to model the 10 spatial dimensions associated with M-Theory, David has embraced a novel mathematical field called octonions. These entities might provide a rigorous mathematical basis for these multiple dimensions using eight unit octonions or dimensional coordinates given as (1, i, j, k, l, li, lj, lk). Each of these indices represents an independent spatial vector dimension. Our present universe requires only four coordinates or three for space and one for time. However, the unusual mathematical construct for octonions is so novel that the order in which mathematical operations are normally performed does not apply to octonions. For example, the sum or difference between two octonions depends upon the order in which they appear in a given mathematical equation. Thus, the arithmetic sum of two octonions, A + B is not the same as B + A, and the product AB is not equal to BA as occurs for matrices. The results obtained from these novel mathematical operations are often surprisingly different and sometimes very novel and insightful. So revised rules for all mathematical operations with octonions must be explored, interpreted and evaluated by David to accommodate the multiple dimensions in M-Theory.

Also, David has been exploring with his students, the consequences if the dimension for time is actually a bi-directional, complex dimension. In this case, time could advance, retreat or stop depending on the algebraic sign on the time dimension. Because of this assumption for the time dimension, many of the bizarre properties associated with quantum mechanics and especially the uncertainly principle become clearer and more rational. It appears that time in the complex plane is actually bi-directional and can move both forward into the future and backward into the past. Such operations with octonions may be essential to develop and evaluate the new properties that occur when ten spatial dimensions are interspersed with the time dimension that are all multidimensional. Such new perspectives and even radical revisions to arithmetical foundations for mathematics may be required as critical steps in confirmation of higher dimensions claimed in M-Theory.

David has contacted his Physics Department Chair at Harvard by telephone and requested that his graduate seminar class be restored to him at the Hospital. David has received approval from the Hospital Administration and requests that the students meet on Wednesday afternoon with David to conduct their advanced seminar sessions. The Department Chair heartily agreed and will inform the students of David's request to reestablish his postgraduate seminars.

David's graduate seminar in hospital (10 Nov Wednesday Morning)

Christa assists David in setting up a small classroom in his Hospital Suite with a white board, various colored markers, and three compact, individual study desks for each of the attendees. These new arrangements in David's Hospital Suite may result in potential patient stress for David from crowding and movement in the small Hospital Suite. Christa must insure that their scheduled classroom time is uninterrupted and productive, but constrained in time duration to the scheduled two hours or less depending upon David's level of exertion during the course session. Furthermore, there must be no telephone or attendant traffic during the session. Any other visitors, or even medical services are excluded unless urgently needed for David's medical care.

Although Christa carefully limits the time and stress demands on David during these class sessions in David's Hospital Suite, she realizes that David must reestablish this important activity to continue his research in physics. The tragic automobile accident has significantly delayed many of the important pending developments that are essential if David is to confirm M-Theory. David is still confident that a viable description of space and time in our universe and the expanded dimensions in a universe surrounding us is demonstrable using M-Theory. To accomplish this, Christa realizes that David needs the support and assistance of his advanced graduate students to provide the essential support, extended research efforts and the iterative interactions he requires to accomplish his objectives.

Before the automobile accident, David supervised these same three students in his graduate research seminar at Harvard. The students consisted of two US doctoral students, Joseph Adams and William Post, and one post-doctoral student named Isaac Haidari from the Technion in Israel. The course agenda planned for the students proposes intense research collaboration with David for a maximum of two hours for each session on Monday through Friday from 10:00 AM to 12:00 PM in David's Hospital Suite. On those occasions when other students and faculty participate, Christa will secure a larger meeting room for the sessions. However, Christa hopes that such expanded sessions will be limited since she realizes that the opportunity for David to overexert during participation is greatly increased then.

Christa assists David in establishing this new seminar agenda with the Chair of the Physics Department at Harvard. A class outline has been provided to the three class attendees who were informed of the meeting times, locations, and seminar expectations for David's In-hospital sessions. On Monday, this seminar team will outline the research direction and development assignments they will pursue during the following weekdays. Then on Friday, they will meet, review their success and failures in their earlier assignments during the past week, and then prepare to return the following Monday with an evaluation of their efforts and the new directions, that are required to continue their work. David will direct the class and make individual student assignments as necessary.

David's three graduate seminar students make their initial visit to his Hospital Suite on Wednesday afternoon. They are warmly greeted by David and Christa. Christa informs the students of the hospital's visitor policy and her requirements for insuring that David does not over exert himself and compromise his medical recovery that is progressing very well. Christa will constrain David's activities during these research sessions if necessary to prevent him from overextending himself.

During their seminar sessions, David is situated in his wheelchair so that he can move about to interact with the students, the white board, and David's computer and printer on the small desk in the Hospital Suite. Christa sits quietly in the Hospital Suite observing the session and monitoring David's activities and especially his movements in the wheelchair as he confronts each student and writes on the white board. David occasionally overstresses himself as he moves back and forth discussing and analyzing the abstract equations they post, modify, and repost in progressive forms on the white board.

However, Christa is tolerant of David's animated participation, within reasonable limits, as David engages with the students. She is delighted to witness David's return to his natural, academic and research element. During these seminar meetings with his students, David becomes completely engrossed, animated, and scientifically engaged with his on-going research in M-Theory and oblivious to external surroundings including Christa. It is only when Christa intercedes and restrains David from attempting to stand from the wheelchair and walk unaided to engage more actively with the students that David acknowledges Christa's presence and reluctantly complies with her strict but gentle restraint. However, with great satisfaction for Christa, she is witnessing David reestablish his passion for physics and his research in M-Theory. Christa muses that her patient is again on the path to confirming M-Theory and hopefully securing a Nobel Prize in physics.

David and Christa's banter (10 Nov Thursday)

Using their individual BlackBerrys, David and Christa can remain in virtual contact with each other even after Christa leaves David's Hospital Suite at the close of her daily shift. In David's apartment, Christa often positions her Blueberry in operating video and audio mode in locations in the kitchen, front room and David's room in the apartment where she and David can converse and interact. David, observing the transformed state of his apartment, frequently compliments Christa on the improvements and orderliness that she has brought to the apartment. The visual access provided by their BlackBerrys also allows David to have virtual access to his research materials in his room. He can observe those items he needs for his research and request that Christa bring the items to the hospital for his use. Even, when Christa leaves the apartment and does her errands in the evening and grocery shopping on Saturdays, David can vicariously accompany Christa using the camera in her BlackBerry. She asks his opinion about selections for food items and even plants and flowers for the apartment to honor his personal input to the apartment. David occasionally requests that Christa bring favorite snacks to the hospital for him. David is particularly fond of roasted, whole almonds, but Christa exercises restraint not to compromise the hospital's food regimen for David's diet.

Christa realizes that these frequent, intimate interactions using these electronic marvels are greatly increasing their relationship and understanding for each other. Furthermore, David is very impressed with Christa's skill in managing her time outside the Hospital. He is learning to appreciate the efforts required to maintain a household. David realizes he is slowly, but effectively discovering the role that a wife and homemaker would provide in his life. For Christa, David's increasing awareness of and appreciation for the duties and contributions of a wife is fully consistent with Christa's plans for her and David.

David's nursing care by Christa (11 Nov Friday)

David realizes that Christa is greatly exceeding her full time, five weekday nursing shift for his exclusive care. In addition to her regular hospital shifts, she often visits him during the weekends and secures materials from his apartment and his office at Harvard to support his scientific research. David has become completely dependent upon her not only for medical support, but also for all those personal needs he cannot perform while hospitalized. Significantly, Christa provides these services with grace and even obvious enjoyment. However, David is more grateful and delighted when Christa is present in his Hospital Suite. During those times, they enjoy their extended conversations, playful banter, discussion about world events and especially probing science and religion. With Christa's sustained and patient effort, David's comprehension and testimony of the Gospel of Jesus Christ is expanding, and the influence of the Holy Ghost upon David is increasing evident to both David and Christa. David and Christa are progressively bonding into an interdependent couple in both secular and spiritual matters with both David's and Christa's approval and effort.

Christ's Atonement for mankind

As Christa and David continue their mutual pursuit of the Gospel, Christa realizes that David must develop a belief and testimony of Christ's Atonement that is essential for salvation and resurrection for humanity. This divine event, the Atonement, is confirmed through the ordinance of the Sacrament. The Sacrament epitomizes the meaning and significance of the emblems that it denotes.

David challenges Christa about the ubiquitous reference to blood throughout Christian Scriptures. He has encountered references to blood in the Old Testament, the New Testament, the Book of Mormon, the Doctrine and Covenants, and the Pearl of Great Price. Christa speaks at length why blood is so significant and emphasized throughout the scriptures. Christa reminds David in their earlier discussions from the Gospel of John, that blood is the host and conveyer of the DNA for mortal men and it was God the Father, through the agency of Christ that our universe is populated with living organisms. DNA is the universal seed for all life on the Earth and is the essential embryo of humanity. DNA carries the complete genetic code of each human being within each unique Genome.

Christa reinforces David's understanding of the significance of human blood and the blood shed by Christ. She reads the following scripture to David to affirm the essential role that Christ's blood provided for the habitation and salvation of humanity.

Acts 17:26
26. *And hath made of one blood all nations of men for to dwell on all the face of the earth, and hath determined the times before appointed, and the bounds of their habitation.*

Christa declares that Mormonism teaches that Adam and Eve were imbued with the immortal DNA genomes of God, the Father and thus Adam and Eve could live forever. However, the transgression of Adam and Eve produced the fall that resulted in the corruption and alteration of this original immortal DNA that was given to Adam as a son of God. Adam's transgression brought about changes in his original genetic code from God and instilled death upon Adam and all his descendants.

Christa states that her nursing education teaches that blood is significant because of its role in transporting DNA and the human Genome through the circulatory system for cell reproduction. However, of all human descendants of Adam, Christ was unique as a being on this earth as the only Begotten Son of the Father in the Flesh and carried both His Father's immortal DNA and His mother's, Mary, mortal DNA. Thus, Christ distinctively possessed the capacity to both suffer mortal death or to retain immortal life within His person.

It was this inimitable, unique capacity within the DNA carried by Christ that allowed Him to perform the Atonement for all humankind. Christ had the power to raise Himself from the bonds of death and this He did upon His resurrection from the dead.

Significance of the sacrament

One of the important prophecies of Isaiah was that not a bone would be broken, but that Christ's blood would heal us. When Christ was crucified and those on the cross were to be taken down, the usual Roman practice was to break the legs of the crucified to insure their death. However, when they came to Christ, it was realized that he was dead, so the Roman Soldier plunged a spear into Christ's side. Blood and fluids gushed out spilling upon the ground. This mass of body fluids over the 33-year lifetime of Christ has deposited molecules of Christ's blood into all the earth's waters. When we are baptized in any water on the Earth, this water contains atoms from his blood.

David's investigates Mormon doctrine using BlackBerrys (11Nov Friday)

Also, most of their intense religious discussions transpire between David and Christa using their BlackBerrys after Christa's regular shift at the Hospital. In the evening in his Hospital Suite while listening to his violin music in the background, David searches the LDS Scriptures using the Mormon based AP that Christa has installed on David's Apple I-Pod. She has provided him with both written and audio copies of all the Mormon Scriptures including the Holy Bible, Book of Mormon, Doctrine and Covenants, and the Pearl of Great Price. In addition, many current Church articles in the ENSIGN and other important Mormon works such as James E. Talmage's "Jesus the Christ" are filed in his I-Pod. David can rapidly search any of these Scriptures or other Mormon works using his topical guide. David finds the topical guide especially effective in scanning all these scriptural references as he seeks answers to specific theological questions that arise as he studies the Gospel. Of particular interest for David are those concepts in Mormon scripture that allude, or reference concepts related to David's interests in M-Theory.

Along with his own investigations, David seeks Christa's interpretation of the scriptures since she is a devout Christian and Mormon. Christa is diligent in answering David's inquiries to the best of her ability and their discussions are revealing and satisfying for David and challenging for Christa. Christa realizes that their BlackBerry sessions in the evening greatly expand their interactions and increase their dependence and love for one another. Most important, these sessions are bringing David to a testimony of the Savior that is Christa's primary goal. David must accept the Gospel if Christa is to marry him.

Christa's weekly Personal Journal for 8-14 Nov 2010

David and I now discuss at length his science and my religion. As a devout Latter Day Saint, I am committed to teaching David the Gospel. Not only because that is what the Savior expects of His Disciples, but also I want David in my life as a Disciple of Christ, a Latter Day Saint and, most emphatically, as my endowed husband. He is diligently studying the LDS Scriptures and believes that they support many of his concepts and properties regarding M-Theory. My faith is based upon my belief in the Savior, His Atonement and love for His children. David's belief in Christ was initiated through his explorations in M-Theory, but the Holy Ghost is now intervening in David's life. It is marvelous to realize that modern science can support many of the beliefs that we espouse and take upon faith as Christians and especially as Mormons.

David has reestablished his graduate class and is teaching his graduate research seminar students at the hospital. This is a very important step, if David is to continue his pursuit of M-Theory. He needs the support of his students to perform much of the research. I will do all I can to insure that he is able to continue his research without endangering his physical recovery. However, I must be careful that he does not overextend himself. He often is caught up in his dialog with the students and overstresses himself in his interactions with the students. I want David to walk out of this Hospital as a strong, healthy, faithful young man and great scientist...Hopefully with me at his side as his wife and eternal companion.

CHAPTER 10 (15 Nov-21 Nov)

Christa will convert David to Mormonism and marry him (19Nov)

David has now become the dominant interest and singular earthly objective in Christa's life. Her thoughts and her passions are fully focused upon David whom she has grown to deeply love and greatly respect. David is not a Mormon, and his Jewish background and scientific experiences are very different from her Christian background and nursing practice. But David is the man that fills Christa's heart and kindles her desires. David is exceedingly intelligent, very handsome, exciting to engage socially and a provocative suitor. Moreover, he is also very kind and thoughtful, and has the qualities and personality that make him an ideal companion…and a good father to the family she seeks to establish. In summary, Christa is profoundly in love with her patient, David ben-Steinmann. Christa's life with David would be challenging but fulfilling, different but complete, intense but exciting, and demanding but satisfying. Living in his apartment, surrounded by his environment, and caring for his daily needs have provided her with tangible experience of what to expect from marriage to this young scientific genius. David is very different from the other men she has known, but it is David who fills her heart, stirs her emotions, and who can provide her with the challenge and opportunity for a full and rewarding life.

Furthermore, David's near-death experience has aroused the Holy Ghost within him and he is rapidly and inevitably, moving toward embracing the Gospel of Jesus Christ. His insatiable drive to merge his scientific research with his NDE should result in David's conversion to the Gospel. Christa is now convinced and committed that David should and will be her husband and eternal companion. Christa has implacable faith that she can and will convert David, this brilliant young Jewish scholar, to the Gospel of Jesus Christ. Christa will become the singular, defining Gospel Missionary in David's life. His knowledge of science and experiences with his NDE and their mutual study of the Scriptures will create a 'new being in Christ' and a husband and eternal companion for Christa.

Christa's weekly Personal Journal for 15-21 Nov 2010

I must declare in my Journal. I now know and testify that I love David and I will make him my husband and eternal companion... with my Heavenly Father's assistance and approval of course. David's understanding and acceptance and testimony of the Gospel of Christ are growing rapidly, and for that, I am truly grateful and blessed. Father in Heaven, please continue to direct and support me in my plans for David and me. David will make a powerful and influential Disciple of Christ; and his conversion to the Gospel will have major impact upon other scientists.

Remarkably, David's increasing testimony of Christ is strongly supported and coupled to his research in M-Theory. David believes that his NDE confirms the existence of higher dimensions and a parallel universe that surrounds our present mortal universe. David now believes that it is the Holy Ghost that is the conduit by which mortal's access and penetrate the veil or membrane for communication with the Father and Son. Through the spiritual medium of the Holy Ghost, mortals may witness the Kingdom of Heaven. Significantly, Mormons receive the Gift of the Holy Ghost upon baptism; and this Gift offers the new Disciple of Christ to access the Father and Son in Prayer and even revelation if warranted by faith and God's approval.

CHAPTER 11 (22 Nov-28 Nov)

David and Christa celebrate Thanksgiving in David's Hospital Suite. (26 Nov)

This Thursday is Thanksgiving and the Hospital will provide a special meal for the patients to accommodate their medical conditions. However, because David's dietary constraints are essentially gone, Christa prepares a special dessert, a rich Jewish butter cake embroidered with dark chocolate and roasted poppy seeds to conclude their Thanksgiving Dinner. She discovered the recipe for this dessert in a Yiddish cookbook that she found in David's apartment. The recipe was marked with hand-written changes made by Rebecca, so Christa believes that David will recognize and enjoy this special dessert. Subtlety, she wants to impress David with her ability to prepare culinary delights as a Jewish wife might for her kosher husband. Even though David does not observe a kosher diet and he will become a Christian...a Mormon, he and Christa can and will honor and maintain his Hebrew cultural heritage and observe appropriate Jewish traditions. Christa realizes that the Savior also observed and honored Jewish traditions during his mortal life as evidenced in the Scriptures. Can Christa do less for her Jewish companion, a literal member of the House of David?

Christa is informed that her mother has died (26 Nov)

When Christa arrives back at the apartment late on Thanksgiving evening, there is a telephone message on the receiver. Christa listens to the heartbreaking message from her mother's Care Center in Utah. The Head Nurse at the Center Informs Christa that her mother died during the afternoon, finally succumbing to her invasive terminal breast cancer. Her mother's passing was expected, and the Head Nurse further states that the death of Christa's mother transpired without undue suffering. The cancer had metastasized, spreading throughout her lymphatic system. The Nurse closes her message to Christa requesting she call the Care Center and provides the staff with instructions regarding embalming and other mortuary services.

Christa immediately calls the Care Center and discusses her options and arrangement for the preparing her mother's body for interment and scheduling the funeral services. After Christa provides the mortuary staff with the required information and funeral arrangements, she calls David and tells him of her mother's death. David is very sympathetic and supportive and expresses his sincere condolences. He tells Christa to make immediate arrangements for a flight to Utah in the morning. She must make all necessary funeral arrangements and close her mother's affairs. David informs Christa that he will cover all the expenses for the funeral. Christa thanks David for his understanding and generosity. She will keep David informed of the services and call him frequently during her absence with her Blackberry.

Christa calls Delta Airlines, secures passage on the first morning flight from Logan International Airport in Boston to the Salt Lake International Airport. She also reserves a rental car in Salt Lake City for local transportation. Christa then calls the Bishop in the Ward that serviced the Care Center. The Bishop agrees to notify the Relief Society President and together they will assist Christa in preparing and conducting the funeral services for her mother.

During the stressful and sorrowful five-day journey and funeral for her mother, Christa travels to Utah, closes her mother's affairs and pays for all outstanding expenses at the Care Center. Christa attends the simple funeral services Monday morning at the Mortuary Chapel. With few attendees, the Funeral Services are very brief and sorrowful for Christa. The Funeral is immediately followed by burial at Wasatch Lawn Cemetery in Salt Lake City where her mother is now interred next to Christa's father. The memorial service is the emotional closure for Christa who realizes she is now without any immediate mortal family members. The loss of her mother, although expected, is particularly distressing since Christa had earnestly hoped that her mother could meet David before her death. Her mother did have a brief conservation with David during their recent telephone exchange.

Christa returns to the apartment from the funeral (30 Nov)

Christa arrives at Logan Airport on Tuesday Evening closing her difficult and emotional trip to Utah. She immediately calls David on her BlackBerry upon landing as he requested. Christa briefly relates the events regarding the funeral, and David expresses his relief for Christa's safe return to Boston. She then takes a taxi to the apartment from the Airport.

Christa is physically exhausted from the stressful trip, but she realizes that David has his final surgery tomorrow. She must be at the Hospital early to prepare him for this last, decisive surgical procedure. David has been supportive and understanding during this difficult time for Christa, and she must now insure the success of this final surgery. David's kindness and sympathy further confirm her love and appreciation for David. He is now the most important mortal in her life. David must become her eternal companion and the center of her new family circle. A family circle composed of an eternal marriage with children for their posterity.

Although Christa knew that her mother's cancer was terminal, the loss of her mother is very traumatic for her. She has no direct family now since both her parents and all of her grandparents are deceased. Christa voices a deep prayer of gratitude for her parents and great sorrow for their deaths. However, she realizes that her mother and father are now together, and their long mortal separation has ended. Christa realizes she must now establish and sustain an eternal family with David as the Patriarch.

As Christa prepares to retire for evening, she opens her journal and ponders over her entries. She then scribes the following words in her journal before her evening prayers and retirement.

Christa's weekly Personal Journal for 22-28 Nov 2010

My mother died on Thanksgiving Day. Her death was inevitable with her terminal cancer and she will no longer suffer the painful effects of the invasive, sinister breast cancer that metastasized and ravaged her body. I know she is now with my father and I am relieved that her suffering has ended. David was understanding and supportive during this difficult time for me. He provided the funds for me to travel to Utah and paid for all the expenses associated with her funeral and closing her medical expenses. The funeral service at the Chapel near the care center was very brief and the Bishop conducted the services. Oh Mother, dearest mother, I miss you greatly, but I comforted with the knowledge of your joyous reunion with my father and your eternal companion.

I observed Thanksgiving last Thursday in David's Hospital Suite. Friends and medical staff joined us. We had roasted turkey, stuffing, yams, cranberries and the other foods that Thanksgiving portrays. I blessed the meal with David's approval and we ate using my Mother's China and her silverware. I prepared a dessert for the dinner. It was a butter cake embroidered with chocolate and poppy seeds. Rosie's Bakery in Cambridge assisted me in properly executing the kosher recipe that Rebecca had used and advising me in preparing the dessert. The Dessert was very successful and I relished David's compliments and pleasure in eating the Dessert.

Significantly, David's testimony of the Savior has increased immensely. I am pleased and grateful for his growth in the Gospel. My love and respect for him enlarges in direct proportion to his strengthening testimony. David, a direct member of the House of David, now recognizes and honors his divine Jewish relative... Christ... the Messiah anticipated by the Jews for millennia. I am certain that David will be baptized, receive the Priesthood, and after our civil marriage, be sealed with me in the Temple. Then my promise to my mother will be honored and we will be an eternal family with my mother and father. Perhaps, David's parents can be sealed with us in the future.

Oh, mother I do miss your mortal presence. I know you suffered these past three years from your aggressive cancer that razed your body. However, your physical suffering is now ended, and you are with my father and both of you are in the tender care of our Savior. But your mortal absence is difficult for me. Your death is a source of continuing sorrow. Perhaps, you were spiritually present with David and me during the Thanksgiving repast, and you witnessed and bestowed your approval for the union of David and me.

CHAPTER 12 (29 Nov-5 Dec)

David's final surgery (1Dec)

David's final surgery is scheduled for Wednesday morning. Dr. Feinstein has assured Christa that this last invasive surgery will complete all of the restorative protocols scheduled and required to repair David's injuries to his hip and leg. Success of this final surgery will accomplish all the required medical procedures to ensure David's full recovery and allow him to be released from the hospital.

Christa arrives early in David's Hospital Suite and she recounts the funeral services for her mother to David. The Relief Society Sisters in the Ward that supported the Care Center for her mother were gracious with their service and generous with their time. Christa is appreciative for all the services and support she received during this difficult time for her. Then she expresses her appreciation for and understanding by David. She is especially grateful for the financial support of the unexpected expenses that resulted from her mother's demise.

Christa tells David that she must now direct all her attention to David and prepare him for his final surgery. David's full recovery and release from the Hospital is essential for David's resumption of his research and teaching at Harvard. Christa then smiles and says his full recovery will sanction their future together. David also expresses that his future plans for Christa and him also depend upon the successful outcome of this last surgical operation.

Christa completes all the preparations that are required for his admission into surgery. She then clasps David's hand and offers a brief oral petition to Heavenly Father for David's welfare. Closing her prayer, Christa assures David that she will again be waiting for his return and continuing her prayers in his behalf. The surgical orderlies soon arrive in David's room and transport him to the surgical suite. Christa accompanies the orderlies to the pre-surgical waiting room. In the waiting room, Christa again clasps David's hand and expresses her love and fervent prayers in his behalf as he is transported into surgery.

Christa returns to David's Hospital Suite and prepares for his return from surgery. She is confident that with the successful conclusion of this surgery, David will be able to leave the hospital walking and able to resume his active role at Harvard University. Christa then muses thoughtfully.

"David will now be able to assume the role of my husband and companion. I know his concern for our future marriage is that he must achieve full recovery from his automobile accident. David has expressed frequently that he will not be a physical burden to anyone, especially for me."

Throughout all David's surgical protocols for correction and mending of his damaged hip and leg, Christa has provided very close medical and physical scrutiny to ensure that each postsurgical recovery is successful, and the serial results of the surgeries will result in the restoration of David's ability to walk and function normally. In addition, to the surgeries, extensive physical therapy has been required and Christa has insured that David dutifully followed and performed each of the therapeutic protocols. David is often reluctant to complete the painful and monotonous exercises and supporting activities associated with the prescribed physical therapies. But Christa is adamant and demands David's compliance. Although, David complains, he realizes that Christa is dedicated to his full recovery. He is grateful for her service and her love and her singular interest in his recovery and welfare. The tacit message between Christa and David is they will marry if David recovers fully from his injuries.

David's return from his final surgery (1Dec)

David's final surgical episode is mercifully short and very successful. David, still anesthetized, is taken to post-surgical recovery where Christa awaits the return of her unconscious patient. Christa is again visited by Dr. Feinstein who informs her that this final surgery was very successful and without complications. He informs Christa that David will be able to leave the

Hospital after recovery from this last procedure. Then with a paternal proclamation, Dr. Feinstein utters that David will depart from Massachusetts General Hospital fully ambulatory.

Dr. Feinstein then candidly tells Christa that David should now be fully confident of his recovery and he will undoubtedly propose marriage to her. Christa warmly thanks him for his extraordinary surgical skills and his delivery of David prepared for a normal life as her husband. Christa gives Dr. Feinstein a warm embrace and expresses her gratitude and blessings for his vital contribution to David's physical recovery and her future with David. Dr. Feinstein, who has assumed a paternal role in Christa's life, wishes her great happiness with David.

Later, David is returned, unconscious, to his own suite from post-surgical recovery. Christa attaches the IV line with Ringers Lactate that deliver the antibiotics prescribed for postsurgical pathogen control. She applies the auxiliary monitoring instruments to measure his blood pressure, blood gases and other vital signs. She insures each of the instruments is functioning properly and that postsurgical recovery is satisfactory. Confident of his full recovery, Christa then tenderly sponges his face, neck and arms. As she views her unconscious patient, she ponders in her mind that with David's full recovery, he will ask her to marry him. As Christa continues her tender administration to his care, David slowly regains consciousness and anxiously calls for Christa.

David awakes from surgery (8Oct Friday)

"Christa, Christa, where are you?"

Christa, at David's bedside, clasps his hand and affirms her presence. She tells him that she has always attended his return from surgery. David smiles at his beautiful caregiver and acknowledges her continuing nursing and declares,

"Yes, Christa, you are the first person I must witness after my surgery. You should now be fully aware of my complete dependence and love you and for my care and wellbeing."

Christa remarks,

"You know I am fully committed to your complete recovery, David." Christa, then with obvious joy and satisfaction, tells David,

"David, Dr. Feinstein has informed me that you will be discharged from the Hospital as soon as you recover from this final surgery. Your leg is now fully restored, and you will leave the Hospital walking on your own. David I am delighted that my prayers in your behalf have been answered."

During the remainder of the week, Christa carefully maintains the auxiliary medical surveillance equipment that monitors David's medical state as she has performed so effectively.

On Friday, the couple close Christa's nursing day and she departs from David's hospital suite, exhausted but joyful.

On Sunday, Christa records in her Journal the following.

Christa's weekly Personal Journal for 29 Nov-5 Dec 2010

David now believes he understands the universe or Kingdom of Heaven in which those beings exist. What is so profound is that David's understanding and testimony of the Gospel now supports his experiences in his NDE, his important research in M-Theory. The Gospel of Jesus Christ as held by the Church of Jesus Christ and Latter Day Saints is embraced by David. These factors are all moving David to achieve a strong and viable testimony and acceptance of the Gospel. For me, this means that David will embrace the Gospel, be baptized and confirmed, received the priesthood and, joyfully, marry me for time and eternity in the Temple.

I am now certain that David is rapidly moving toward marriage with me. I love him dearly, but he must be baptized and join the Church for us to wed. My desire and commitment to marry a worthy member and Disciple of Christ and be sealed in the Temple is unaltered and absolute.

However, I am convinced that David will become a Disciple of Christ and David will become my eternal companion. Although David is Jewish, and Judaism denies that their prophesized Messiah was Christ, David is now under the compelling influence of the Holy Ghost. Christ was also Jewish, and had the constant companionship of the Holy Ghost since He was the only begotten Son of God the Father. I am confident that with the sustained influence of the Holy Ghost and my prayers and efforts, David will become a Disciple of Christ and my husband.

CHAPTER 13 (6 Dec-12 Dec)

David's petite mal seizure (10 Dec Friday)

Christa is in David's apartment preparing to retire late Friday evening when she receives a call from the Staff nurse on David's floor. The Staff nurse tells Christa,

"Your patient, David Steinmann, has experienced an episodic epileptic seizure. The seizure was classified as a 'petite mal' episode, but the seizure has now terminated. David is presently resting; and he is calm and sedated. However, because of his extensive and intrusive surgeries, he remains under close supervision in an Emergency Recovery Room at the Hospital. As you aware, he cannot tolerate his previous anti-seizure medication. Had the seizure escalated to a 'grand mal,' the consequences might have proven fatal."

Christa anxiously declares that she will return to the Hospital immediately. She rapidly dresses and drives David's long-term rental car to the employee parking lot at the Hospital. The information clerk at Emergency Admissions Office greets her and tells Christa that David is in Emergency Recovery Room 1011 on the ground floor. Christa enters the emergency wing and is informed by the attendant that David is resting but sedated from the demurral administered by the floor physician to prevent possible injury to his right leg from lingering spasms. Although the seizure was not severe, David did dislodge the restrains for his leg and the duty physician anesthetized David to avoid possible disruption of the skeletal alignment made during his recent surgery.

Christa informs the attendant that she will remain the evening at David's bedside to monitor his signs and insure there is no reoccurrence of the seizure or possible damage to his leg. Christa enters David's Hospital Suite and moves a chair next to his bedside. She carefully examines the unconscious man who now commands her time and attention. Christa checks the medical instruments monitoring David's vital signs. His blood gases are normal; oxygen levels at 98% and respiration rate and blood pressures are satisfactory. She insures that the fixtures constraining David's right leg are firmly attached and competent. She then settles into the chair next to David and prepares for the all-night vigil.

Christa action at the hospital (10 Dec late Friday)

Gazing upon David's ashen face, Christa realizes she is deeply in love with this young Jewish scholar who now dominates her thoughts and life. His image and being are constantly present in Christa's mind. Although they are worlds apart in education and background, David has captured her heart and soul. Indeed, her conscious actions and thoughts are now centered in David, his recovery, his future, and, hopefully, their life together as husband and wife.

While Christa ponders David's dominance in her life, David suddenly arouses and unconsciously pulls on his constrained leg. Christa stands and quickly, but gently restrains his motion to ensure that he does dislodge his tethered leg. David remains heavily sedated and unaware of Christa's presence. As she holds her hand on his chest and leg to restrain any further involuntary reflexes, Christa is moved by her love for this young man who is now the dominant individual in her life. She slowly bows her head, and gently clasps David's head between her hands and kisses David's forehead. Her embrace and kiss is a physical manifestation of her concern and devotion for this unconscious patient. David is her sole object of interest and passion. She smiles inwardly and vows that she will make this unlikely suitor her husband and companion for time and eternity. Ending her kiss, Christa whispers softly,

"You are the man, David Steinmann, who will fulfill the life-long dream of this young Mormon woman for an eternal companion."

Surveillance camera records Christa's kiss (10 Dec Friday)

While this tender and ardent interlude is occurring, Christa is unaware that the surveillance camera in David's Emergency Suite has captured Christa's embrace and kiss upon the unconscious David. Christa has forgotten that surveillance cameras are dispersed throughout the Hospital and monitor most areas, especially those in the emergency ward where patient welfare is paramount and continuous tracking and medical telemetry are essential.

The display on the surveillance camera in David's Emergency Suite has exhibited Christa's actions to members of the nursing staff on the floor. Significantly, those who have observed Christa's act include the Head Nurse for the Emergency Ward.

Head nurse confronts Christa (10 Dec Friday)

The Head Nurse for the Emergency Ward, upon observing this scene, quickly enters the Emergency Suite and hostilely confronts Christa. She declares that Christa's actions are unacceptable for a nurse at Massachusetts General Hospital and a violation of hospital policy. Nurses are not to display such affection towards their patients whether they are conscious or unconscious. The Head nurse angrily declares that she will inform the Hospital Supervisor of this unacceptable act by Christa and she will undoubtedly be discharged from the Hospital staff. Christa vainly attempts to explain her actions and the exclusive and intensive care she has provided for David during his extended hospital commitment. However, the Head nurse ignores Christa's explanations and orders her to leave David's Emergency Suite immediately. The Head nurse will assign another nurse to monitor this patient while he remains in the Emergency Ward. A distraught and frightened Christa complies with this demand and returns to David's apartment for the remainder of the weekend.

Upon her return to the apartment in the early hours of Saturday morning, Christa remains sleepless. She has great anxiety and uncertainty regarding her actions. Will her spontaneous act of affection towards David result in her discharge from David's care and perhaps even the Hospital Staff? Can such a simple and innocent gesture of compassion end her aspirations and hopes that Christa has held for David? Could such a spontaneous, harmless event terminate Christa's quest to woe and win the hand of David. It is inconceivable that her innocent display of affection could destroy her hopes of union with David. Christa broods over the events of that Friday evening with David in turmoil over the weekend. She closes Saturday evening with a fervent prayer that this incident will not end her relationship with David and her position at the Hospital.

Christa calls the Emergency Ward on Saturday to determine David's condition, but she is informed that David is still in his emergency room and is unavailable to be disturbed upon orders of the Head Nurse for the Emergency Ward.

Sunday for Christa (11 Dec Saturday)

Sunday is also an anxious day for Christa. Christa calls the Emergency Ward and is informed that David has been returned to his hospital suite. The threat by the Head Nurse hangs over her posing ominous consequences. She attends church, but her mind is troubled. As she takes the Sacrament, Christa silently utters repentance for her spontaneous act. Christa does not actively participate in her Gospel Doctrine class or her Relief Society Class. Her mind is occupied about incident in the Emergency Ward. Returning to the apartment after church, she inadvertently picks up the telephone to call her mother and discuss the problem, but then realizes her mother is deceased now and that this previous avenue of safety and peace of mind is gone from her. Her mother is with her departed father and Christa must cope with this crisis alone. She yearns to share the problem with David, but she is unable to speak with him. She must resolve this threat alone.

After her sparse evening meal and her poignant prayers to her Heavenly Father, she moves into her bed, takes her journal from the nightstand, and pens the following words:

Christa's weekly Personal Journal for 6-12 Dec 2010

David experienced a petite mal seizure. I went to the Hospital to insure his recovery and prevent possible harm to his injured leg. While I was watching him, I was aroused by my love for David and kissed him on the forehead, as he lay unconscious on his bed. However, the Head Nurse in the Emergency Ward observed my act and accosted me. She has threatened me with discharge from the Hospital. Can such a simple act of love for David result in loss of David as my patient and my nursing position at the Hospital? Please, Father in Heaven, intervene in my behalf. Preserve me from my indiscretion. My fate is in Your hands, and I pray that You will resolve this mistake and sustain my nursing assignment with David. I love David and I cannot lose him.

My dear mother has now departed this life and her years of suffering are ended. But I do so miss her presence and counsel. Her death from terminal cancer was inevitable, and a blessing, but my loss is still profound. I miss her mortal influence in my life. I can now only share my great anxiety over this innocent act of affection for David with my Heavenly Father and pray for His intervention and resolution.

CHAPTER 14 (13 Dec-19 Dec)

Hospital administrator dismisses charge (13 Dec Monday)

Upon Christa's return to the Hospital the following Monday morning, she is informed by her Nursing Floor Administrator that she must meet with the Hospital Administrator before assuming her nursing duties. Christa is terrified and obediently goes to the Administrator's office. After waiting in the Administrators reception room for several anxious minutes, the Administrator comes to the reception room and tells Christa to enter his office and be seated. As Christa enters the office, the Administrator immediately observes the anxiety and fear that shrouds this competent but distressed young nurse. Unknown to Christa, the Administrator is familiar with Christa's nursing assignment, and Dr. Feinstein has informed him of the special relationship that exists between David Steinmann and Christa.

The Administrator officially relates his understanding of the Friday night's event in David's recovery suite reading from the Head Nurse's written report. Christa is frozen with fear and tears appear in her soft pale blue eyes as the Administrator concludes the formal report. Then with a warm, disarming smile, the Administrator remarks,

"Christa, you are one of my finest and most dedicated and competent nurses at this Hospital. But you should exercise discreteness in your future relationship with David Steinmann, especially in a Hospital suite where the surveillance cameras are active. I also have a supplementary report from Dr. Feinstein who corrects the report from the Head Nurse regarding your special relationship with patient, Dr. David Steinmann."

Then with a disarming response, the Administrator continues his assessment of the events regarding Christa's actions,

"Of course, Ms. Olsen, it is contrary to hospital policy that the attending staff osculate their patients even if such acts enhance their medical recovery. However, Christa, I am aware of the excellent care that

195

you have provided our famous patient and the close relationship you have with each other."

The Administrator now stands, smiles warmly, and offers Christa his hand.

"Christa, you will continue to administer to David Steinmann's nursing needs. He has fully recovered from his epileptic seizure and is now in his regular hospital suite. Please return to his room and his singular nursing care."

Christa's face immediately brightens, and she immediately stands and grasps the hand of the Administrator in gratitude and apologizes for her action with David. The Administrator then grins and candidly discloses,

"Massachusetts General Hospital would be delighted to later announce that a female Massachusetts General Hospital member of our general nursing staff has become the bride of Harvard's famous physics patient currently resident at the Hospital."

The Administrator concludes his words.

"Nevertheless, it is unlikely that all our single male patients will marry their attending female nurses, despite the excellent care they receive at Massachusetts General Hospital."

Christa is buoyant with gratitude to the Administrator and departs his Office greatly relieved. As she joyfully returns to David's Hospital Suite and resumes her regular nursing assignment, she silently expresses gratitude to the Lord for His quiet but compelling mediation in her behalf and Dr. Feinstein's intervention.

David's inexperience in human relations (13 Dec Monday)

David has learned earlier of the rumors of Christa's ardent kiss while he was unconscious in the Emergency Ward. Therefore, when Christa enters his Hospital Suite, he immediately calls Christa to his bedside and teasingly tells Christa that he was informed that she had kissed him while he was unconscious in the Emergency Ward. Christa is embarrassed by

David's words and apologizes to him. David, however, responds with joyful fervor and vigorously declares,

"I am very sorry that I was not awake. Then the kiss would have lasted much longer from overt and sustained action on my part."

David then banters further saying,

"Christa, let us repeat that kiss and embrace again; I know I will enjoy it more when I am conscious and can visually observe and embrace my beautiful caretaker. That act of ardor on your part will result in a greatly sustained duration of passion on my part."

Christa gently scolds David and responds,

"No more kissing between us in this Hospital, I have fretted and suffered greatly over the incident. My weekend was dominated by my anxiety over my act."

This successfully concludes the petite mal incident that threatened the maturing relationship between the Mormon Nurse and the Jewish physicist.

David's admission to Christa (13 Dec Monday)

Later, following a candid discussion with Christa about physical affection between men and women, David confides that he has never romantically kissed any woman in the past. He quickly asserts that he is not without passion and Christa's beauty arouses his ardor. But he has no previous experience in courtship or close, amorous relationships with women. He never experienced his mother offering physical affection or even conspicuous warmth towards him or his sister Rebecca. Their childhood was a rigid regimen of chores, school, and religious studies and observance of Jewish traditions. These traditions all occurred without demonstrable affection or even verbal approval. Only stern discipline and strict schedules were expected and demanded from family members in the Steinmann household. David then pauses and says reflectively,

"My mother did discipline me to be very diligent and successful in my academic pursuits. She is probably responsible for my profound commitment to science and effectiveness in research. But that strict and austere environment deprived me of familial love and inexperience with true love and compassion for others."

David quickly asserts that he is not unacquainted with the emotional side of affection. He certainly is aware of Christa's physical beauty and attraction, but what does an embrace or warm greeting between husband and wife, a mother and child, and even friends truly entail and signify?

David is embarrassed to admit his ignorance and inexperience, but his past observance of Christa, her warmth and spontaneous and open display of tenderness and her emotional grace and ease with others has aroused his concern. David then ruefully declares,

"I am inexperienced with and ignorant of close and loving relations among people, particularly within a family and between a husband and wife."

David repeats his earlier assertion saying,

"Again, don't believe I am without emotions. My interest in you is very active and intense; especially as I have watched, you move and work. You are a lovely young woman, a work of art in motion. I am continually fascinated with your beauty and demeanor. I am physically attracted to you, but can I learn to fully respect and honor you as a child of God with an immortal soul, mind, and heart. These emotive issues have troubled me, and I need your support and understanding, Christa. If we married, I want to be the husband and companion and the father to our children that you expect and deserve?"

Christa is moved by David's concern and declaration of intention to marry Christa. She responds thoughtfully,

"David, I observed your mother's interaction with you only once when she confronted you here in the hospital after Rebecca's death. I am truly sorry that she did not exhibit the warmth and love that can and should exist between a mother and child during their childhood. Depravation of that critical bonding is an unfortunate loss during the early formative years of a child. However, that loss can be overcome, and I believe that we can establish that essential human quality of love and concern that can and should exist among spouses, children and other family members. I was blessed to have a mother and father who deeply loved one another and their only child, me. Each member of my family was fully aware of the love and concern we had for our family members. Family hardships, illnesses, and even disagreements would temporarily strain relations among us. However, there was never doubt that each family member was very important, unconditionally loved, and valued above any divisive issue that might arise. That binding, familial love was particularly important and severely tested when my father was killed while landing his aircraft aboard the U.S. Navy Aircraft Carrier, Enterprise that was deployed in the Pacific. The aircraft arresting cable failed to restrain his F-18 Super Hornet, and it plunged overboard off the bow of the flight deck. The ship rammed the downed aircraft and killed my father. I still grieve over his death and lament over his absence in my life. The loss of my father was horrific, but the love and support with my mother sustained us through the difficult times. Now she too is deceased. Nevertheless, my parent's unconditional love for each other and for me, still sustains me."

David sees grief in Christa's pale blue eyes as she recalls the tragic accident that removed her mortal father from her during her life and the recent death of her mother. After a long pause as she regains her composure, Christa returns to David's admission of inexperience with close family relations and emphasizes,

"But do not fear my dear David; I have sufficient experience with familial relations and love for you, that we would be successful in our marriage, and our love and respect for each other would provide the basis for compassion and love for each other and for our children."

David, learning of Christa's emotional recovery and maturity from this tragic incident in her childhood, gently responds,

"You are a remarkable woman whose maturity and wisdom match your beauty. I know that you can teach me to develop that rapport, respect, and intimate love that should exist between a husband and wife in a functional, loving family."

Christa assessment of David (14Dec)

After her shift at the Hospital while traveling on the MBTA to return to David's apartment, Christa ponders her fervent discussion with David about his inadequate feelings of love and compassion for others. The depravation of family love and support during his childhood can be overcome, and Christa is certain she can correct David's childhood depravation. Besides possessing a brilliant mind, David is also kind, loving and arouses Christa's love and provokes her passion. Furthermore, David is now truly a follower of Christ as well as a moral, virtuous man. He has arrived at his faith in Christ by means of a different path than most Disciples of Christ, but his faith is strong and secure. Science has dominated David's life, and science is the principal driving force for his conversion to the Gospel of Jesus Christ. Christa realizes that true science also testifies of Christ and David has discovered that relationship with the Divine. David is now convinced that Christa's Christ is the Messiah that Jews have sought for millennia. Remarkably, David has achieved part of his testimony from experiences and evidence gained through his NDE. David realizes that our universe and other possible multiverses confirm that God the Father is the Supreme Being, and acceptance of and discipleship with Christ is essential for salvation and exaltation. Christa also realizes that David's conversion and testimony may attract other unbelieving scientific minds to ponder and accept Christianity. Inspired men of science may now contemplate and accept the Gospel of Jesus Christ as discovered by David and promulgated by His Church, the Church of Jesus Christ of Latter Day Saints.

Attention by David's colleagues for Christa (15 Dec)

During a hospital visit the following day by one of David's single male faculty colleagues at Harvard, Christa is present and performing her usual nursing responsibilities. While discussing M-Theory research, David's colleague intently observes Christa and inquires about her marriage status. He sees that she does not have a wedding ring, and he informs David that his nurse is very attractive. If she is single, he wants David to introduce Christa to him.

David is suddenly confronted with competition for Christa's attention. The sudden realization that he could lose her to another suitor surprises and alarms him. He knows very well that Christa is, indeed, a very attractive and desirable young woman; and he has no claim upon her, at least not yet. David realizes he must act, and act quickly and decisively.

David tells his colleague that Christa is soon to be engaged and married to a man here in the hospital. Hearing David's words, his colleague comments that her suitor is fortunate and abandons further effort for a formal introduction to Christa. David has successfully diverted the issue, and his colleague soon leaves after he and David conclude their mutual research discussions. However, David now recognizes that Christa is exposed to the attention of other men and is a choice prospect for courtship and even marriage to someone else. He has taken Christa's commitment to him for granted, and that would be a disaster for him if he were to lose her now.

Later, while Christa is attending to regular nursing duties in David's Hospital Suite, he confronts Christa and subtlety suggests she should consider looking at engagement rings. An engagement ring will inform other men that she is unavailable for dating and courtship. Christa smiles at David's comments and inquires if David is actually jealous of other men and their attention for her. David responds firmly,

"Yes, and I am very possessive as well as jealous when you are the object of other men's attention. I have important plans for us, and I will not let them be thwarted by other potential suitors."

With a confirming smile, Christa responds warmly,

"I look forward to learning of these mutual plans."

Christa then departs David's room to secure his medical prescriptions for the evening. As she walks to the hospital pharmacy, she silently muses to herself at David's response, and she realizes that David is rapidly moving toward their future marriage.

Christa's engagement ring (17Dec Friday)

Although their courtship has evolved steadily considering the intensity and concentration of their continuous relationship and interaction within the hospital and their respective BlackBerry communications outside the Hospital, this is the singular and memorable day that David decisively proposes marriage to Christa. This extraordinary event begins unexpectedly for Christa as she prepares to leave David for the weekend on Friday evening. After a full, exhaustive shift of nursing duties, Christa approaches David's bed and says,

"David, this will be a very busy weekend for me. I must close outstanding issues regarding my mother's funeral services and her final financial obligations. In addition, besides restoring the apartment that I have neglected these past several weeks since my mother's death, I have laundry to do, bills to pay, groceries to buy, and some special errands for our apartment tenants, particularly the dear widow on the second floor I have befriended. Also, I must respond to your mother's recent letters requesting closure of Rebecca's personal affairs. I will need your advice to reply to her petition in a satisfactory and gentle manner. I do not want you to incur more resentment from her. We must seek to console her over Rebecca's death and gain her acceptance of Rebecca's tragic accident."

As David listens to Christa's duties, mostly in his behalf, he realizes what Christa does for him and how fully he has come to depend upon her for all his needs and obligations during his extended medical convalescence. He has delayed seeking Christa as his wife because of his concern that he would not recover his full physical strength and agility. However, he is now fully confident that his injured leg will adequately heal, and he will be strong and fully ambulatory. He realizes he can now assertively entice Christa into his life as his companion with no physical handicaps. Furthermore, he knows that he deeply loves Christa and wants and needs her in his life…permanently. She is beautiful, charming, and competent and will be an ideal companion for him. He cannot resume his future life without her. Most important, he is learning how to love another person, especially this unique and beautiful woman who has altered his life, introduced him to the Gospel of Jesus Christ, and secured his physical recovery as well as aroused his passions and fulfilled his desires for a companion.

David's marriage proposal (17Dec Friday)

So while Christa is at his bedside saying farewell, David reaches out and gently clasps Christa's hands in his and declares tenderly,

"Christa, I have pondered for some time how to approach you with my desires and plans for both of us. Now that I am assured that I will regain full mobility and agility with my leg, I can now confidently voice my petition to you. I realize that I am inept at expressing emotion and my profound love for you, but let me try. These last three months together with you daily have transformed my life and confirmed my need for a companion…a choice, beautiful, eternal companion. The loss of Rebecca and the many duties she performed for me without adequate appreciation on my part are now very evident by your service and unconditional support for me. However, more important, you have not only filled that void and restored my leg and insured my mobility, but you have introduced me to the tender, unfamiliar feelings of affection, support, respect and dependence upon a companionship of compassion and love for one another. Initially, I was fascinated and attracted by your physical beauty, your femininity, and your grace and intellect. You still stir my passions and your purity and virtue arouse my feelings and desires for you. But, in addition to my love for you, I have become completely dependent upon you for my ability to cope with life outside of my commitment to science. I know that I am deficient in dealing effectively with others, negligent in paying bills and meeting obligations, and simply functioning as a responsible human being in a hectic, demanding, secular world. I took Rebecca for granted since she was my sister, and I provided for her physical support."

"You, Christa, have not only provided and satisfied all those needs but aroused within me the need I have for your continuous companionship and my own spiritual fulfillment together with you. I have come to share your faith and testimony in God, and His Son Jesus Christ and His marvelous Universe. My near death experience, our intimate religious discussions and investigations, and your profound presence and impact in my life have opened my eyes to the joy that a man and woman can experience in this life with each other and their mutual love and relationship with the Savior."

As David continues his proposal, Christa's own emotions and deep love for David swells within her breast and are now evident in her face and demeanor. Then David declares,

"Christa, I need you permanently in my life. You will provide the purpose and joy that I seek and need to fulfill in my life. You would greatly honor me, if you would become my wife... my companion not only for this life, but also for the life to come...an eternal, exalted life. Yes, I do believe that your Savior...Jesus Christ is the Messiah that was sought by my people, the Jews, but rejected by them. When Christ returns to the earth again, His people, the Jews, who remain on the earth then will recognize and acknowledge Him as their long sought Messiah."

"Christ is real and alive and has and will secure the salvation and immortality for all mankind. He died for our transgressions and His Atonement will absolve all the sins and wrongs present in this life. Through baptism in His name and obedience to His commandments, we can become exalted heirs in God's Kingdom. Christa, the Holy Ghost has witnessed this testimony to me."

Christa's is now euphoric as David then resolutely declares,

"Christa, I will be baptized and confirmed a member of the Church of Jesus Christ if you will join me in marriage. We can be married later in your Temple and sealed together after our civil marriage here in the hospital. Christa, my love, will you marry me?"

Tears of joy now fill Christa's eyes as she realizes that her cherished goals and intense desires for their union have now been fulfilled. She understands that David loves her as intensely and completely as she loves him. He has proposed marriage and will satisfy all her essential needs for their marriage and eternal union. He will be baptized and confirmed a member of the Church, and they can later be sealed together in the Temple after David receives the Melchizedek Priesthood. Christa's promise to her mother to marry a worthy holder of the Priesthood can now be honored through her celestial marriage to David Steinmann.

Christa can no longer contain her joy and exhilaration and she reaches forward with both of her hands and warmly embraces David's face and kisses him ardently. Christa's tears of joy course down her silky-smooth cheeks. After their sustained caress she utters with excitement and passion,

"Oh David, my love, you have granted my greatest desire...to become your wife...and eternal companion will be an honor and great joy for me. I have hoped and prayed for this outcome. I have loved you since the first day you were admitted to Massachusetts General Hospital. Your presence in my life has been the driving desire for our marriage and eternal companionship. Your proposal for our marriage and promise to embrace the

Gospel of Jesus Christ will also please my deceased father and mother. David, I will make our marital union wonderful, filled with love, passion and charity; fruitful with children; and we will be sealed together...forever. I will fully support you in your commitment to physics and your goals to disclose the true nature of this marvelous Universe that God has prepared for His children. Your research will assist others in science to realize the wonders and blessings that our Father in Heaven has prepared for His children."

Christa, with tears of joy still streaming down her moisten checks and her hands embracing David's face, Christa then charges provocatively,

"Dr. Steinmann, I expect a Nobel Prize Medal to honor your work and reside in a prominent and honored place in our home."

David is now equally stirred with emotion as Christa continues,

"David, I too share your passion for disclosing the truths of this universe and God's Kingdom. The marvelous life that God the Father and His Son, Jesus Christ have prepared for His Disciples now and in the future with Him and His Father in the Celestial Kingdom will be ours forever. I testify that God is delighted that you have discovered and disclosed many truths about His Kingdom and its properties. Our marriage and union will enrich and sustain your research in achieving that realization."

David smiles at his beautiful fiancée, and laments that he cannot fully embrace and caress his 'beautiful bride to be.' However, they will share those exciting and sacred physical and emotional moments later after their marriage.

Christa's engagement ring fits perfectly (17 Dec)

David then requests that Christa secure the appropriate notification to the world to confirm his love and intentions to make her his wife as he says,

"Christa, my love, I have a safe deposit box at the First National Bank in Cambridge. The key to the safety deposit box is with the keys to the apartment and the key is stamped with 'First National Bank of Boston' on the key face. My safety deposit box number is the largest, four-digit prime number, '9991'. In the safety deposit box is my paternal Grandmother's diamond wedding ring. My grandfather gave that ring to my Grandmother after he rescued her from certain death when she was imprisoned at Auswitz during the Holocaust. He placed that ring upon her finger later when he married her. She wore that ring every day of her life after their marriage until her demise. When my Grandmother died, my Grandfather gave me my Grandmother's ring for my future wife. The ring was to be a serial heritage for the Steinmann family and an emblem of my grandparent's profound love and commitment to each other. Now that binding circle of gold will become the external symbol of my eternal love and commitment to you, Christa. My Grandmother's ring has great sentimental value for me, and your acceptance of the ring would fulfill my solemn promise to my Grandfather. Christa, my love, please secure this emblematic ring from the safety deposit box and have the solitary blue-white diamond set in a mounting that pleases you and will fit your slender finger."

Christa is euphoric and declares to David she will secure the ring and bring it to David for their binding wedding engagement.

Christa departs Massachusetts General Hospital this Friday evening with complete fulfillment and utter joy. Her hopes and plans for their future marriage have been realized. David has proposed marriage and she has joyously accepted David's proposal with all of her heart and love. Christa's fervent prayers for their union will now be fulfilled . She deeply loves David and David's love for her will now be confirmed with the wedding ring that belonged to his Grandmother. As Christa walks to the MBTA station, she is ecstatic and filled with joy gratitude to the Lord. As she walks to the MBTA station, Christa recites in her mind the joyous declaration,

"David ben-Steinmann will now become my husband and my eternal companion."

Christa recaps this Sunday (18 Dec Saturday)

Over the weekend after many frequent emotional and passionate conversations and mutual prayers with David using their BlackBerrys, Christa pauses on Sunday evening before retiring and reflects on the marvelous events of the past week. Christa softly vocalizes on the fateful blessings she has received.

"What an incredible transformation a week has wrought in my life. Last Sunday I feared that I would lose David forever and be discharged from the Hospital. This week, David has proposed marriage to me and has agreed to be baptized, receive the priesthood, and take me to the temple to be sealed forever as his eternal companion."

Christa ponders these events and then declares softly,

"Your abundant blessings in my life, Heavenly Father, are marvelous and humbling. These blessings transcend my greatest hopes and desires. To be the wife of David ben-Steinmann, world-renowned theoretical physicist who will convert to Christianity and Mormonism is incredible, but true and wonderful. This marvelous blessing could have occurred only with Christ's approbation and the divine influence of the Holy Ghost upon David."

Christa then gazes upon the bare ring finger of her left hand and solemnly promises,

"I will wear David's Grandmother's diamond ring with honor and pride. I will learn more about Jewish family traditions and honor these in my life with David. Indeed, many of these traditions were observed by the Savior, and He and His family honored these traditions and observed hallowed Jewish rituals. We will celebrate the Seder, the Feast of the Passover, and observe and keep the Christian Sabbath in our home. We will visit the Temple frequently, the Temples of the Church of Jesus Christ that are built and designed after the Temple where Christ expounded His Gospel during the meridian of time. What joy and blessings will be ours together as man and wife…as eternal companions…hopefully with children to bless our union."

With great joy and gratitude after completing her evening prayer with David using their mutual BlackBerrys, Christa bids David a pleasant night and peaceful rest with her love and anticipation for their formal engagement and marriage. She will be in his Hospital Suite on Monday morning after the bank opens. She will secure and proudly deliver the revered Steinmann engagement ring that she will ask David to place on the ring finger of her left hand to announce and confirm their engagement.

To document the realization of this blessed, culminating event in her life, Christa removes her Journal from the nightstand and writes the following declaration in her personal journal.

Christa's weekly Personal Journal for 13-19 Dec 2010

David has proposed marriage and I have joyfully accepted. I have hoped and prayed for this event and now it has come to pass. I will marry David and we two will become one flesh as the Scriptures declare. Marvelously, David will marry me, has agreed to be baptized, confirmed, and will receive the Melchizedek Priesthood. We will later be married and sealed in the Temple and receive our endowments.

The wedding ring that was worn by David's Grandmother will be a sacred token of our mutual love and commitment for one another. I am ecstatic with joy, greatly blessed and will be honored to display the ring that is an important emblem of David's family and his heritage and now my heritage.

All my dreams, hopes and aspirations for us have now been realized. I will be the companion of a worthy Disciple of Christ who is indeed, a literal, ancestral member of the House of David. Christ, the Savior of Mankind was a member of the House of David. I am truly honored and grateful to share this ancestry. Thank you, Heavenly Father, for this magnificent blessing in my life.

CHAPTER 15 (20 Dec-26 Dec)

Christa accepts David's engagement ring (20Dec)

Before her morning shift the following Monday at Massachusetts General Hospital, Christa eagerly enters the First National Bank in Cambridge upon opening. She anxiously signs the access card, and immediately goes to David's safety deposit box. She identifies the safety deposit box to the bank clerk who uses his master key and David's key to free the box for removal from the vault repository. Waiting for the clerk to leave, Christa then opens the box and carefully secures the small jewelry box with David's Grandmother's ring. With great anticipation, she slowly opens the small black jewelry box.

Inside the tiny, velvet lined jewelry box is an exquisite yellow gold ring with a large single, multi-carat, brilliant white diamond. Christa's long, slender fingers gently remove and embrace the ring and feel its texture. She ponders how David's Grandmother must have felt when she received this stunning ring from David's Grandfather. Then with quasi-reverence, Christa slowly slides the ring onto the third finger of her left hand. To her surprise and immense delight, the ring is a perfect fit. She lifts her hand to the overhead light and the multi-facets of the diamond illuminate and shimmer through the primary spectral colors as she slowly moves the ring across the bright xenon lights in the vault's ceiling. David's Grandmother's wedding ring is a precise fit and a perfect bond for their engagement and marriage. Christa pauses, smiles, and ponders the feelings she will experience when David places the ring on her left hand signifying their engagement.

Christa then carefully returns the ring to the jewelry box, safely secures the box in her purse, and impatiently departs for her shift at the Hospital. It is difficult for Christa to contain her excitement as she boards MBTA for Massachusetts General Hospital and David's Hospital Suite.

When Christa enters David's Hospital Suite, she is exuberant and buoyant with joy. She quickly approaches his bed and her joy and excitement is obvious and unconstrained. David smiles and inquires if she found the ring. Christa cannot contain her excitement and utters with delight as she stands at his bedside.

"David, dearest David, the ring is a perfect fit. No alteration in size or change in setting is necessary or desired. I will be honored if you will place your sacred Grandmother's precious wedding ring on my finger. Her ring, her original, unaltered wedding ring, I will wear her ring with great honor and consummate joy as your wife."

Christa solemnly delivers the jewelry box to David. He opens the box, carefully removes the ring and gently takes Christa's left hand in his left hand. Then with his right hand, David gently slides the ring onto Christa's slender ring finger. This is the simple confirming act of their engagement. Christa bows her head and gives David a brief, but ardent kiss. Then Christa declares as she exhibits the solitary diamond ring now displayed on her finger,

"David, see...the ring is a perfect fit."

David beams at his beautiful, future bride and completes her words,

"Yes Christa, my love, the ring is a perfect fit for a perfect bride and a perfect marriage."

David and Christa close this prenuptial scene with a warm embrace and a bonding kiss of love.

Christmas with David (25 Dec)

Saturday is Christmas and although not a scheduled workday for Christa at the Hospital, she arrives at 8:00 AM to greet David and spend Christmas Day with him. When she enters the suite, David is sitting in his wheel chair with his computer in his lap. She quickly lays her Christmas packages for David and others at the hospital on the counter and David sets his laptop computer on his table and closes the cover. He greets Christa with a smile and a brief kiss and then says,

"I am so glad you are here. Christmas would be a dismal day without my beautiful nurse and my fiancée with me. With you present, we can now celebrate the birth of the Savior together. I have called a caterer and they will deliver a special Christmas brunch for us at 11:00 AM. I hope you had a light breakfast or perhaps you fasted this morning. Regardless, the brunch will include some favorite Jewish menus that I hope you will enjoy. After all, you will soon assume and command the role of the Jewish matriarch in the David ben-Steinmann household."

Christa responds earnestly,

"David, I want to learn to prepare some Jewish menus, particularly items that you enjoy, and I can successfully prepare. I am curious, should we begin to amend our diet now to be Kosher? I have been reading about what Jewish families are expected to eat. I now have recipes for Challah, Hummus, Mandelbrot, and Strudel. What should I learn to prepare to delight my handsome Jewish husband?"

David answers,

"Do not fret, my love, my dietary habits are fully Americanized. As an adopted and confirmed Yankee, I consume all pork products including ham and even well cooked bacon, which is unthinkable for a 'Kosher' Jew. I consume dairy products and meat served from the same plate, and I violate most other kosher laws. However, I realize that too much cholesterol is not wise. I will simply await and enjoy whatever my beautiful, Christian bride prepares for our repasts. I know that whatever food you prepare in our home will be delicious and nourishing."

Christa responds,

"Actually, David, I do want to expand my culinary skills by preparing some of the Jewish recipes that I have encountered in my reading and the Kosher Shops I have visited in Boston. I want to become a role model for Jewish cuisine, although I will remain a Mormon and a Christian. Can I do that and at the same time honor my husband's heritage? David, the Jewish

213

people have marvelous and ancient customs and traditions. I will even try to learn the Hebrew language. I want and expect that our children will learn English and Hebrew and Arabic and perhaps some of the other languages you know. I have learned that 'boker tov' is the salutary greeting for 'good day.' I would love to have you read the Old Testament to me in Hebrew. Most Christian Scripture is devoted to the descendants of Abraham who was the Patriarch for the Twelve Tribes of Israel."

"David, did you know that as a Mormon we hold firmly to the history and destiny of the Jews as the chosen people of God. We teach and fully believe in the return of the twelve tribes of Israel to Jerusalem. We recognize and honor the lineage of Abraham, Isaac, and Jacob. I am considered an adopted member of the Tribe of Ephraim, one of the sons of Joseph who was sold into Egypt by his brothers. I learned of my lineage regarding the House of Israel from my Patriarchal Blessing that I received while I was a nursing student at BYU in Provo, Utah. David, you must be a member of the Tribe of Judah that was the Savior's direct mortal lineage. It will be interesting to have you receive a Patriarchal Blessing. I wonder what the Patriarch will declare in your Patriarchal Blessing regarding your lineage and Priesthood Blessings."

Christa sits next to David's wheel chair and they hold hands warmly conversing. Frequently, Christa lifts her left hand to admire her engagement ring. David comments that the ring is apt emblem of his love for Christa. He asks Christa to shop for a matching wedding band that will complement her engagement ring. Christa smiles and declares simply that she will have mutual wedding bands to exchange during their marriage ceremony.

There is a small Christmas tree in David's Hospital Suite and Christa stands and places the Christmas gifts she brought under the tree. She remarks with a smile that there are many gifts to both David and Christa from the Hospital staff. David informs Christa, that the staff placed small trees in most of the patient's room and then distributed the gifts that were delivered to the nurses' stations.

Many of the Hospital Staff who have served David in some way during hospitalization came by David's Hospital Suite and wished the couple a Merry Christmas. In addition, everyone congratulated us on our pending marriage here in the Hospital. The nurses greatly admired Christa's stunning engagement ring that Christa displayed with pride.

A very unusual and greatly appreciated gift is that which dealt with the detailed, multipage invoice for medical services for David's medical expenses at Massachusetts General Hospital. The final tally of charges was over a million dollars and would have caused Christa great financial concern. David's medical insurance coverage through Harvard University financed only about half of the cost, and the remainder was their personal debt. However, when Christa opened the Christmas card from the Hospital Administrator, which was also endorsed by Dr. Feinstein, there was a statement that the Hospital would dismiss the remaining expenses for David's hospitalization and services. Christa was deeply grateful to the Hospital staff for its great kindness and service to David. Since Christa will now be responsible for financial affairs in this marriage, she was concerned if they could manage these very large medical expenses. With the Hospital statement was a financial statement from Dr. Feinstein that dismissed all his surgical charges along with those of his medical team and other staff.

Then Christa opened Dr. Feinstein's separate Christmas card to David and her that has the simple statement.

"Best wishes to our brilliant, young Jewish patient and scientist and his lovely Mormon nurse, soon to become a married couple, Shalom and 'Shalom Aleichem' (peace be upon you)."

On the back of Dr. Feinstein's card were the following inscriptions in English and Hebrew.

To the unlikely, but very choice, couple soon to be married.

Joseph
Feinstein

יוֹסֵף

בָּרוּךְ יְהֹוָה לִי אֱלֹהֵינוּ מֶלֶךְ הָעוֹלָם
שֶׁהֶחֱיָנוּ לְקִיְמָנוּ לְהַגִּיעָנוּ לַזְּמַן הַזֶּה.

Curious of its content, Christa asked David to articulate the words for her in Hebrew and then translate the words into English. David examined the Hebrew words and then voiced them with solemn elocution.

"Baruch Atah Adonai, Eloheinu Melech Ha'Olam, Sh'hecheyanu, V'Kiyemanu, V'Higianu LaZman HaZeh."

David then reverentially translated the sacred Hebraic words into English for Christa.

"Praised are You, the Eternal One our God, Ruler of the Cosmos, who has kept us alive, sustained us, and enabled us to reach this moment."

Tears welled up in Christa eyes, and David squeezes her hand gently as they exchanged a simple kiss of affection. After this solemn exchange Christa turns to David and says,

"I want to ensure that Hebrew traditions as well as Christmas and Easter observances become an important part of our lives together. I will honor and respect the Jewish ancestry that my husband and our Savior both bear. Our family will be blessed with this rich heritage and our children will be reared in a home filled with love, with the Savior's teachings and the Spirit of the Holy Ghost."

Christmas continues (25 Dec)

For Christmas, David gave Christa a beautiful bouquet of flowers…Orchids, her favorite. Another gift from David was a beautiful mother of pearl necklace. Among the gifts Christa gave David was a rare,

216

limited edition of a mathematical text in German that he requested for his research. Christa finally found this rare tome as a used text in Europe. Only after extensive searching, did she find a small Austrian library that had the book and would reluctantly sell it to her. The book was badly worn, yet the library wanted $1100 for this very rare copy. However, David desperately needed the book, so Christa offered $1000 and acquired the expensive volume.

Christa briefly scanned the book while wrapping it for David's Christmas present. The book contained an obscure mathematical treatise on "octonions." David told Christa that octonions are the essential mathematical vocabulary for multidimensional spaces in higher mathematics. Christa hoped that the book was useful to David for his work in M-Theory. David was excited upon receiving the rare book and informed Christa that the book would be very valuable for his research and resolving many mathematical questions. He examined the text frequently during the day, commenting that there were many unique theorems and mathematical insights that were significant for his research.

The brunch that David ordered for delivery at 11:00 AM was delivered on time. The caterers for the meal brought a small folding table, set it with a green, holly tablecloth that said Merry Christmas, and even lighted the seven-candle menorah that provided the perfect Jewish décor for our Christmas feast. The dishes were served, properly warmed, on the table from insulated food carriers. The main course was roasted turkey breast with yams, cranberry topping, snowcap peas, warm dinner rolls and wassail.

Wassail is a cider drink that contains fruit juices, cinnamon, cloves and other spices. This Christmas drink was a tradition in the Olsen Family and David ordered the drink especially for me. Christa told David that she fondly remembered the first stanza of the traditional carol associated with Gloucestershire Wassail dating back to the Middle Ages. Christa's dear father would sing the Welsh strain during Christmas especially with his navy aviator comrades when they visited our home. Their sonnet was as follows:

> *Wassail! Wassail! all over the town,*
> *Our toast it is white and our ale it is brown;*
> *Our bowl it is made of the white maple tree;*
> *With the Wassailing bowl, we'll drink to thee.*

After the meal, David and Christa each sat in comfortable chairs and spoke of our future together. Of course, the honeymoon and their scheduled trip to Israel and Jerusalem were the highlights of conversation for me. David outlined a brief, tentative itinerary of the sights he had planned for the couple to visit. Their Honeymoon trip to the Holy Land was unanticipated and exciting for Christa. David could see Christa's joy in planning and visiting that ancient, sacred land that hosted the Son of God during His mortal sojourn.

Later after many personal visits from others from the Hospital and Harvard, David spent the remainder of this Christmas Day in his wheelchair and Christa sat closely by his side. He completed some calculations on his computer and Christa read from her Scriptures. Later Christa reread the Christmas cards they had received and then softly voiced the immortal passages from Luke announcing the mortal birth of Christ declared in Luke, second chapter 2, versus 8 through 11.

> *And there were in the same country shepherds abiding in the field, keeping watch over their flock by night. And, lo, the angel of the Lord came upon them, and the glory of the Lord shown round about them; and they were sore afraid. And the angel said unto them, Fear not; for, behold, I bring you good tidings of great joy, which shall be to all people. For unto you is born this day in the city of David a Savior, which is Christ the Lord."*

David is now able to walk short distances unaided, although Christa accompanies him to ensure there are no risks from his ambulatory excursions. The last surgical procedure occurred earlier in December and was very successful. Dr. Feinstein assured me that David can be discharged in early January and he should be fully ambulatory then with no or minimal walking restrictions.

At 5:00 PM, Christa could see that David was tiring from the long Christmas day, so she bid him farewell. They exchange a kiss and a long embrace, and then Christa left for the apartment. Christa told David I would call him on the BlackBerry and let him know she returned safely to the apartment. Christa drove back to the apartment in the long-term loan vehicle provided by David's auto insurance company as part of the settlement since the accident in October. Christa preferred to use public transportation, but the holiday schedules for transportation are very limited and inconvenient. Christa had driven little in Boston traffic and she must renew my driving skills since David wants her to drive our rental vehicle in Israel. They will tour most of the sites in Israel during their visit...Their honeymoon in the Holy Land. Christa must accommodate to driving in a foreign land, especially in Israel.

Christa's Sunday after Christmas (25 Dec)

On Sunday as Christa partook of the Sacrament, she pondered in my mind the recent events in her life. It is remarkable that less than three months ago, Christa did not know David and he was not part of her life or future plans or grand aspirations. Only the Lord could have wrought such blessings from a terrible accident that took the life of David's sister Rebecca and nearly ended David's ability to walk. Now, within twelve weeks, Christa had found her husband and eternal companion, a brilliant young Jewish scientist, who she had, or more correctly, the Lord had brought into the Gospel of Jesus Christ and sealed that union by their marriage. Christa pondered the unlikely, but marvelous couple they would become. Christa could not have planned or accomplished a more improbable and remarkable union. Indeed, these events and their fruition in Christa's life were beyond her limited powers and vision. Only the Lord could have forged such a union of two diverse people. The Hands of the Lord are very evident in Christa life and David's, and their blessings were marvelous and unfathomable. Christmas was a remarkable day for both of Christa and David. It was a wonderful and sacred experience. Christa deeply loves David.

Later in the evening after praying with David on their mutual BlackBerry phones and after settling into her bed, Christa removes her Journal from the nightstand and writes.

Christa's weekly Personal Journal for 20-26 Dec 2010

What a marvelous and blessed week this has been. I am engaged to be married and my husband is David ben-Steinmann, the famous, endowed professor at Harvard. Besides being a unique and remarkable scientist, he is also a choice young man. I love him for many reasons; he is honest, kind, good, and loves me as much as I love him. However, of greater significance, he will join the Church and become a Disciple of Christ, which is the most important and essential requirement for our union.

These three months as David's nurse, close companion, vicarious roommate and confidant, and now his finance have changed and blessed my life. I wish my dear mother could have actually met David. I know she would have accepted and loved him as I do. In truth, I believe my mother and father who are now deceased do know and love David as I do. David fills my heart and soul, and we are so good for each other. I truly believe I am the ideal wife for this giant in the esoteric world of physics and M-Theory. He loves his science and research and marvelously, they have given him a testimony of Christ. To have David become a believer in the Savior through scripture study, prayer, science and his pursuit of M-Theory and his NDE is miraculous and the work of the Holy Ghost. That M-Theory and God's Kingdom are compatible and even linked is a strong testimony for David and for me.

We celebrated Christmas together in David's Hospital Suite. This was the first Christmas that David has celebrated as a believer and disciple of Christ. His Jewish family, especially his mother, Sariah, did not condone the observance of Christmas or even acknowledge this Universal

Holiday in her home. Nevertheless, David and I enjoyed the special day and I will make it an important tradition in our lives together.

I gave David several gifts including a technical book he wanted for his research. David had me purchase my wedding dress...a beautiful, white chiffon gown with tailored sleeves, a short train, white slippers and silk stockings. I told him that the marriage gown I really want is that sacred temple gown I will wear in the Temple when we are married there and sealed for time and eternity. David chides me and says I may not want him as a companion eternally, but I counter his jesting remark and tell him he will always be mine...forever...even after death separates us. That is the promise that the Gospel of Jesus Christ offers to its endowed disciples.

CHAPTER 16 (27 Dec-2 Jan)

David and Christa's wedding plans

After considerable planning and discussions in the Hospital and using our BlackBerrys in the evenings, the couple schedules the Wedding in the Reception Room at Massachusetts General Hospital for 10:00 AM on 10 January 2010. The Bishop of Christa's Ward agreed to perform the Civil Ceremony and visited David in the Hospital to interview him. Bishop Ulrich is a Research Scientist at MIT and is familiar with David's research and position at Harvard. The Bishop looked forward to meeting and visiting with David. Christa sent out a few Wedding Announcements on Monday evening after her shift at the Hospital with David. She obtained a list of those persons David wished to invite on his behalf. Christa asked if she should send an invitation to his mother in Israel or perhaps call her on the telephone. David declined the offer and informed Christa he wanted nothing to disturb that important day in their lives. Perhaps, they could inform David's mother of their marriage during their honeymoon trip to Israel.

The last week of the year passed very rapidly for Christa since there were so many affairs and arrangements she had to complete. Besides her own dress, she had to obtain suitable attire for David. She realized his antipathy for formal male garb, but Christa insisted that she wished to display my handsome new husband; and clothes were important for that ceremony.

New Year's events for David and Christa

Friday is New Year's Eve and Christa told David she would stay at the hospital past midnight with him to welcome the New Year. She sat most of the evening near David while he worked on his research and Christa attended to their mutual mail, bills, and other perfunctory tasks. David sat in a chair with his computer in his lap and oblivious to the time. At 11:45 PM Christa turned to David and said,

"Can you spare a few moments from your research and hold my hand as we celebrate the New Year and our New Life together. On the tenth day of January in the year 2010 of our Lord, I will become Mrs. David ben-Steinmann. David, I am so happy and grateful for you and our new life together."

David sets his computer on his table and closes the cover. He greets Christa with a loving smile and affirms tenderly,

"I am also so glad you are in my life, sweetheart. This New Year will begin our exciting, rewarding life together. The last three months have altered both of our lives and especially mine with the discovery and winning of my bride and eternal companion. I fully agree with your observation. God has blessed us immensely, especially me by gaining the hand and love of a beautiful bride and companion, a choice daughter of God who complements and delights me in every way. You fill the many deficiencies in my character, my lack of social experience and my need for love and compassion. I now realize how empty and unfulfilling my life would be without you. What a unique blessing you are and a gift from God in my life."

The couple sit together hand in hand whispering their mutual love for one another. At midnight, they are surrounded by the distant merrymaking and the muted sounds of the New Year that echo in the environment of the Hospital.

After midnight and the beginning of a new year in the life of David and Christa Steinmann, the couple tenderly kiss, and Christa departs the hospital. Christa drives David's long-term rental car back to the apartment. Upon arriving at the apartment, Christa is joyfully exhausted and quickly pens the following entry into her Journal dwell

---

Christa's weekly Personal Journal for 27 Dec-2 Jan 2011

We will be married on Monday, the 10th day of January in 2010 in the Massachusetts General Hospital Reception Room. My bishop in the Cambridge Ward will perform the ceremony. Then David will take me to Israel for our honeymoon. I am overwhelmed with my blessings. Oh, Father in Heaven...I am greatly blessed and grateful to thee. Marriage to a righteous companion, a literal descendent of the Tribe of Judah in the House of Israel, a honeymoon in Israel, and a visit to the Sacred City of Jerusalem that Christ walked will greatly bless my life. These remarkable events are extraordinary gifts and abundant blessings from my Heavenly Father.

My promise to my mother to only marry a man who would be worthy to enter the Temple and become a righteous son of God and my eternal companion can now be fulfilled.

Finally, I have resolved, with David's full approval, that we will begin our family immediately on our honeymoon as the Lord expects us to do. I want a child, hopefully a son for David.

CHAPTER 17 (3 Jan-9 Jan)

David and Christa prepare for their wedding

With their wedding scheduled for next Monday morning, David and Christa spend their weekdays in his Hospital Suite planning and preparing for the wedding. David is able to move unaided about the Hospital Suite and within the Hospital corridors with relative ease. However, Christa is concerned with his movement and travel so that he does not over exert himself or risk falling and damaging the numerous fragile sutures that cover his right leg. Fortunately, David is a rapid healer and has experienced no infection or trauma from his many surgical episodes. Christa attributes this success to David's superb medical staff and especially Dr. Feinstein's remarkable surgical skills. Christa is grateful to Dr. Feinstein for restoring David's leg and returning David to full ambulatory status. Christa realizes that David would not have asked her to become his wife if he were not fully mobile and physically strong. During their short, but intense courtship that began that first Saturday when David met Christa in his Hospital Suite as his assigned nurse, David would only consider marriage to Christa if he were fully ambulatory and strong. However, Christa realizes her prayers for his recovery have been amply answered by the Savior.

Christa contemplates the events that have brought David to her

Christa muses thoughtfully,

"I am still overwhelmed with the events that have wrought me to this sacred event in my life. Although our courtship has been short, only three amazing months, David and I have submerged ourselves in each other's lives. I believe the intensity and depth of our relationship, discussions of religion and science, the importance of the Savior in both of our lives and my appreciation and understanding of David's research and its bearing on our marvelous universe, perhaps universes, which our Heavenly Father has designed and prepared for his children. David's testimony of the Savior is strong and viable, and I will continue to expand and promote his full discipleship with Christ."

David's final medical examination

On Wednesday, David had his final medical examination. The team of examiners was headed by Dr. Feinstein and a medical board. The examination included detailed assessment of the extensive historical set of CT scans of his injured leg as he progressed with the very successful surgical procedures. The team was unanimous in their assessment that David's surgeries and physical therapy had been remarkably successful. With care and exercise, he will recover full strength and mobility of his leg. Indeed, within a year there should be no symptoms of the major injuries he suffered in the severe automobile accident that took the life of his sister Rebecca. David will be released from Massachusetts General Hospital following his marriage to Christa Olsen.

The assessment of the examining medical team was received by David and Christa with great joy and appreciation. To David, this was his confirmation and his absolute requirement for their marriage. He had often told Christa that he would not marry her unless he could function as a strong, vigorous male and husband. He would not encumber Christa with a handicapped scientist, as he quaintly alluded to himself, much to Christa's displeasure.

Christa takes her Journal from the nightstand and vocally rehearses her entry.

"We will be married in the Hospital and then later leave for our Honeymoon to Israel. Our honeymoon plans are confirmed, and they will include an extended trip to Israel, with concentration on Jerusalem. David did this to please me and satisfy the long yearning that I have held since childhood to see the land where Christ dwelt and preached and healed. Not only has all mankind been blessed by the Gospel, but also Christ's atonement insures that all would live again."

Christa then pens into her Journal the following:

Christa's weekly Personal Journal for 3-9 Jan 2011

The marriage is tomorrow on Monday, 10 January 2011 at 10:00 AM Eastern Standard Time. I have arranged for my Bishop in the Cambridge First Ward to perform the Ceremony. The Bishop, who is a member of the Harvard Administration and knows David, interviewed him earlier in the week regarding our marriage. The Bishop assured himself that David has a firm testimony of Jesus Christ as the Only Begotten Son of God the Father. The Civil Marriage License for the State of Massachusetts is prepared and ready for our signatures after the ceremony. I will become the legal wife of David ben-Steinmann and my married name will become Christa Olsen Steinmann. Moreover, I have David's promise that we will be sealed together as an eternal couple in a Temple Ceremony. My sacred goal has now been realized with the confirmation of the Holy Ghost.

David is ambulatory now and he stands and walks using his cane only occasionally. David has received a positive medical prognosis by Dr. Feinstein and David will be released from the hospital on Tuesday. I have been abundantly blessed and the joy of being David's wife and companion is the realization of my fervent prayers. I love David and he loves me.

Remarkably, we fly to Israel on Thursday, only three days after the marriage, for our Honeymoon in the Holy Land. David has acquired airline tickets on El Al for the flight from Boston's Logan Airport to de Gaulle Airport in France and then to Ben-Gurion Airport in Israel. We have our passports and I have already packed our luggage for traveling.

I am so excited for these life-altering events. I hope David is strong and has recovered sufficiently for this stressful travel itinerary. However, I will carefully monitor his activities and ensure that he is not overstressed. I have successfully served as his nurse, and caretaker, and now I will serve as his wife.

Oh, to actually witness and experience Jerusalem and Bethlehem...my childhood dreams will be realized. To walk the land the Savior strode two millennia earlier, and to do so with my Jewish husband, a descendent of the Savior of the World...what incredible, abundant blessings flow from my Heavenly Father.

CHAPTER 18 (10 Jan-16 Jan)

Marriage conducted at hospital (10Jan)

Bishop Richard Ulrich from the Cambridge Ward in Massachusetts will perform the marriage ceremony in a Hospital Administrator's Reception Room of the Hospital. Because of the limited size of the Room and hospital and Massachusetts Fire Code restrictions, the number of guests will be limited. However, the attendees at the wedding are notable. The President of Harvard, the Hospital Administrator, David's colleagues and graduate students and some of Christa's close hospital associates will attend. David's mother, Sariah, will not attend. David did not want Christa to invite David's mother, although Christa is disappointed and pressed David to inform her of the wedding. Perhaps with time she will accept Christa and recognize and accept David's Christian Bride.

David has insisted that he will stand throughout the ceremony, without a cane. He will also visit with the guests and properly display his beautiful bride before all the wedding guests.

Christa's wedding gown is a simple, pure white full-length satin gown; but her perfect feminine stature and personal beauty present her as a celestial bride in David's eyes. However, David does regret that Christa's silky blonde hair is not configured with her usual ponytail tied gently to drape over her back. He expresses that he has grown accustomed to her normal nursing appearance that he expected and revered. He fondly remarks to Christa that that is one of Christa's remarkable beauty traits that he first observed upon gaining consciousness from his Emergency Ward ordeal from his auto accident. David will miss this inimitable coiffure signature at the wedding that he treasures for his new bride. To accommodate David, Christa promises him that she will don her bundled tress throughout their honeymoon after the wedding.

David uncomfortably wears a white shirt and black tie with a rented black suit. The black, tie-laced patent leather shoes are tight fitting and awkward for walking. Such formal attire for David is an uncomfortable and unique event. However, Christa insists that he dress appropriately if only to please his bride and the wedding guests. David, as usual, generously complies with her wishes and acknowledges that this wedding day is Christa's singular day of Jubilee.

David discharged from Hospital (11Jan)

Following the wedding, the nuptial celebrations and the well wishes, and farewells to the many hospital staff that corrected his serious injuries and restored his ability to walk, David is discharged from the Hospital. His physical recovery from the extensive surgeries and demanding physical therapy sessions have been remarkably successful. However, he is cautioned by Dr. Feinstein that he should avoid extended periods of standing and long distances of travel by foot. He is also advised to use a cane for safety and avoid excessive lifting and extended stair climbing. Finally, as a new groom, he must forego carrying his beautiful new bride across the threshold of their wedding chamber, wherever that may be.

David and Christa return to David's apartment for the first few days of their marriage. Christa is very insistent that they begin their conjugal union with prayer and thanksgiving for David's recovery from his severe injuries in his automobile accident. This they do with David first offering their vocal prayer followed by Christa's supplementary oblation. Their tender and compassionate oblations provide the sacred initiation for consummating their marriage and union.

David is delighted with his beautiful bride. She fulfills all of his expectations for marriage and companionship. She is passionate, loving, warm, sincere...a physical and spiritual companion in every way. To his added delight, Christa is also a fine cook, an excellent homemaker and a frugal, financial manager.

David and Christa to honeymoon in Jerusalem (12Jan)

Among the many connubial delights and post wedding plans that David and Christa experience in their apartment in Belmont is their pending honeymoon. The couple carefully plans their itinerary for their visit to Israel. Christa has dreamed of a trip to Jerusalem, the Holy Land, where Christ walked and delivered His Gospel of Good News to all mankind two millennia earlier. David knows that their honeymoon in the 'City of David' established by David's progenitor, King David, is Christa's profound aspiration.

For Christa to be there with her Jewish husband, a literal member of the tribe of Judah and a descendent of Christ, will be a matchless, spiritual experience as a Disciple of Christ. Furthermore, with David's familiarity with the land, kinship with the people, and fluency with all the indigenous languages in the region including Hebrew, Arabic and Yiddish will make Christa's honeymoon unique and forever memorable.

Before their departure for Israel, David completes necessary financial matters including changing his term life insurance beneficiary from Rebecca to his new wife, Christa. Also, because of his concern for his long-term health and Christa's need for financial security, he increases the death benefit for Christa to $2 million. Christa challenges him over this large annuity and increased expense as excessive and unnecessary, but David is resolute. He must insure Christa's financial independence upon his death. David cautions Christa that sudden, fatal epilepsy is a dominant reality in David's life. Christa reluctantly complies with David's request.

David plans their itinerary to Israel (12Jan)

As part of their planning, David says he has family and many friends living throughout Israel. He wants to confer with his past scientific colleagues, both students and faculty he knew while studying at the Technion in Haifa. Of course, David's mother also lives in Jerusalem, and he may have to broach the issue of his new wife if his mother becomes aware of their presence in Israel. David is concerned how his mother will treat his Christian bride if they were to visit her. Although he has not informed his mother of his marriage to Christa, he is certain Sariah is aware of his marriage through her many Rabbinical Jewish contacts in the United States. Some of them must have informed Sariah regarding David's marriage to Christa. David realizes many of his Jewish colleagues are also as provincial and biased against gentiles as is his mother. However, David dismisses this disagreeable possibility for the time being and focuses on his attention on his honeymoon with Christa in Israel.

David has also planned a short visit to CERN, Switzerland to confer with his experimental colleagues there working on the Large Hadron Collider. He has developed additional, theoretical insights that might support experimental confirmation of M-Theory. The Weisenhausern Trust Directors have strongly suggested, even expect that David include a productive visit to the LHC in conjunction with his trip to Israel.

Side trip to the LHC in CERN and visit in Switzerland

As David prepares their final itinerary for their travels, he casually asks Christa,

"Christa, would you entertain a short visit to CERN, Switzerland after our arrival and stopover in Paris the day before? I need to confer with some of my colleagues there at the LHC Facility. Such a short diversion in our transit would be very helpful to support my research."

Christa beams with delight with David's offer and joyously declares,

"David, an evening layover in Paris and a visit to Switzerland would be unexpected and wonderful. What more could a young bride wish for her honeymoon than additional visits to such romantic and exceptional sites as we travel to Israel. A trip to Paris is the dream of every young bride. However, for me, Switzerland is even more exciting and special. My maternal grandmother was born and reared in Basel, Switzerland. She cared for me as a child when my father and mother had short recreation and relaxation visits, called R&R, together in Hawaii. She often spoke nostalgically of her home in Switzerland and her love for the Swiss Alps. However, David, can we afford such an extended trip? I know that you are anxious to continue your research and I have tried to be frugal with our household budget."

David counters,

"Christa, next to my decision to marry you; the next best decision I made was to have you manage our finances and household expenses. My own budgeting practice was capricious and poorly controlled. Frankly, I have little interest in personal finances and I let Rebecca handled my affairs. However, you have placed us on firm financial ground and stable control of our income and expenses. You have been very frugal and efficient, and you have managed our household budget with skill and thrift. I have no concerns over expenses for our honeymoon."

"Furthermore, sweetheart, I was informed by the Weisenhausern Trust Directors that they would give us an additional stipend if we went to CERN, and I conferred with the Staff at the LHC. I believe the Trustees are anxious over my commitment to my research after meeting you at the Dinner Party at Harvard last evening. One of Directors said, after visiting with my beautiful young bride, that I might have lost my solitary, consuming passion for physics. Christa, do you think I have?"

Christa embraces David and emphatically responds,

"No David, I know you love me deeply and passionately, but I will not allow you to divert or diminish your research in physics and especially your goal to secure experimental evidence for M-Theory. I am fully aware of and supportive of your pursuits in science and your commitment to your research. Professor Steinmann, I honor and embrace your love of physics. Sharing of your work in M-Theory has also aroused my interest and commitment. Furthermore, I want a Nobel Medal to honor your efforts and achievements as much as the Weisenhausern and Harvard Trustees do."

Then after their spontaneous embrace closes with a sustained kiss, Christa continues,

"I married you for countless reasons, and a compelling reason is your desire to unveil the Kingdom of our Father in Heaven and His Son, Jesus Christ. Your testimony that the Holy Ghost is the conduit through which mortals pierce the membrane or veil that separates this mortal world from God's Kingdom and accesses our Heavenly Father is very profound and I believe is true."

David completes their travel itinerary and presents the result for his new bride. Christa is delighted with the plans and anticipates their visit to each of the cities and sites David has identified.

David and Christa begin their honeymoon travel (12Jan)

Christa has carefully packed all the clothes and personal items they will need for the trip. The couple travels by taxi to JFK Airport and boards the late evening, Boeing 747 El Al aircraft bound for Paris. After arrival the next morning at De Gaulle Airport in Paris, they spend the day casually walking the Champs Elsie and River Seine ending with a Paris Taxi ride for brief visit to the Louvre. Christa is ecstatic to see the major work's, particularly witnessing the Mona Lisa by De Vinci. Thoroughly exhausted, the couple takes a taxi to their hotel for the evening.

The next morning they fly to the Geneva Airport, and take a 15-minute taxi ride to CERN. Founded in 1954, the CERN Laboratory sits astride the Franco–Swiss border near Geneva. David tells Christa that CERN is one of Europe's first joint scientific ventures and now has 20 Member States including the United States.

David spends the three days conferring with the staff at the LHC while Christa visits the sites around Geneva and takes the celebrity tour of the CERN Laboratory. Christa is greatly impressed with the size and scientific equipment at the Large Hadron Collider. Following David's conference in CERN, they then board a regional airline and travel to Ben-Gurion Airport in Israel.

David and Christa's travel plans in Israel (12Jan)

Upon arrival in Israel, they will rent a vehicle and travel to Tel Aviv, Caesarea, Haifa, Nazareth, the Sea of Galilee, the Mount of Beatitudes, Capernaum, Jerusalem, Bethlehem, Masada, and Beer Shiva. From Beer Shiva they will then return to Ben-Gurion Airport for their return trip to the United States. Christa will do all of the driving in Israel because of David's concern over a potential seizure. Christa quickly acclimates to driving in Israel as she did in Switzerland.

To Christa's delight and strong suggestions, they will include visits to many of the villages and historical sites where Christ was born, lived, traveled, and preached His Gospel and Atoned for all mankind. Christ is now the essential Figure in the religious and secular lives of David and Christa as they visit the Holy Land.

Arrival at Tel Aviv (13Jan)

Upon landing at Ben-Gurion Airport and passage through Visa Control, David rents a compact car for the entire stay in Israel. They travel to the Marriot Hotel in Tel Aviv where they have reservations for their first evening in Israel. After unpacking, the couple, after a long embrace, withdraws to the dining room at the hotel where they enjoy a romantic candlelight dinner. Since this is Christa's first trip to the Holy Land, she insists that they enjoy a Honeymoon nuptial Dinner their first day in Israel.

For their dinner in the Marriott Dining Room, the newlywed couple dress formally for the occasion at Christa's insistence. Christa is elegantly outfitted in her new wool worsted wardrobe that she purchased in Boston before their departure. To please David, Christa's long hair is elegantly bound in a high ponytail with a black silk ribbon that accentuates her golden blonde tresses. David fondly watches Christa fashion his favorite coiffure that captured David's attention upon awakening the first day in his hospital room.

Christa believes that David is handsome in his white dinner jacket, black matching trousers and a blue button-down open collared that she also acquired in Boston before their departure. Christa allows David to omit the usual necktie that she knows he loathes wearing.

As they arrive in the Marriott Dining Room and are seated by the Maître Dee, David, to Christa's delight, orders their meal in Hebrew from their Jewish server. As David adroitly glides through the full coarse venue for their dinner, he unwittingly orders a red wine for their Salmon entree. Christa recognizes his Hebrew choice for their beverage when the waiter shows David the wine menu for final selection. Christa gently clasps David's hand and prompts him that they do not consume alcohol even for this special occasion. Christa reminds David that he will be baptized at the BYU Center this Saturday...the Jewish Sabbath. She does not want him to require repentance before his baptism so she softly consuls him,

"David, Mormons do not consume alcohol...remember the Word of Wisdom."

The waiter is confused by Christa's words. David, recalling the New Testament, counters that Christ turned water into wine for the wedding guests at the feast in Cana as His first miracle. She and David are now guests at their own wedding feast, so can David do less? However, Christa cleverly undermines his reasoning and responds,

"David, you may order water, and if you can change the water to wine, then we will both have wine with our dinner."

David relinquishes his request and responds,

"Touché my love, you have again reminded me that you are my equal in wit and wisdom."

She smiles as David lifts her left ring hand and places a gentle kiss on her slender palm. With her hand in his, they mutually admire the ancestral diamond ring that encircles Christa's slender left finger and is the physical symbol of their marriage and union. David completes the orders for their evening meal with wine omitted from the order. David and Christa then enjoy the full course dinner and culminate the meal with a favorite Jewish dessert that David orders. To Christa's delight, David orders Butter cake that Christa prepared for David on Thanksgiving in the Hospital.

After a romantic and elegant dinner, the newlyweds retire to their apartment for their first evening in Jerusalem.

David baptized and receives Melchizedek priesthood at BYU Jerusalem Center (15Jan)

The next morning David checks out from the Marriot Hotel. The couple packs their luggage in the rental car, and David and Christa begin their driving tour through Israel. Christa drives north on Israeli Highway 5 North and their first stop is Caesarea, the ancient capital built by Herod on the Mediterranean shore. Herod was the Great and cruel ruler of Israel during Christ's birth. They visit the great arena that faces the Mediterranean and the couple stroll the ruins together. David locates the stone artifact among the ruins that bears the image of Pontius Pilot. David translates the engravings for Christa and comments that this stone is a major archeological artifact attesting to Governor Pilot's reign in Jerusalem during Christ's Earthly ministry.

Then they travel to Haifa, which is the major shipping harbor for international trade for Israel. This modern, urban center is the home of the Technion, which David attended as an undergraduate student in physics. As they stroll hand in hand, David escorts Christa though some of the classrooms and other buildings that David frequented as a student. As they tour the buildings and grounds, David encounters familiar administrators and faculty to whom David proudly introduces his new bride.

Their next stop is a drive south to the Mount of Beatitudes and Capernaum that are important historical Christian sites immediately north of the Sea of Galilee. The couple visits the ruins at Capernaum and observes artifacts and remnant structures thought to have been present during Christ's frequent visits to that famous Biblical site.

Christa asks David if they can walk along the north shore of the Sea of Galilee. Then Christa drives the rental car to the shore where the Monastery resides. With great expectation, as they arrive at the beach, Christa removes her traveling shoes, asks David to carry her shoes, and she walks bare footed along the sandy shore. Christa tells David that this inland, fresh water sea was a special location for Peter who was a Fisherman in this water. She reminds David that many significant Gospel passages occurred near and on the Sea of Galilee. Christa strolls the shoreline chasing the fluctuating waterline while David walks abreast of her on the shore side and carries her shoes. The gentle surf occasionally covers Christa's bare feet and she wistfully remarks that the Savior also walked this same shore. As the couple leisurely stroll, they encounter a fisherman casting his lure into the Sea of Galilee and David and Christa pause at this pastoral scene. Observing the intent seaman who is oblivious to David and Christa, David playfully asks Christa,

"Would you have married me if I were a fisherman? Peter, the great Jewish Apostle, was a fisherman living here in Capernaum with his wife."

Christa answers in the affirmative with her usual provocative smile stating,

"Yes, my dear David."

Christa then counters with her own declaration,

"I would have also married you if you were a carpenter. Christ, the Savior of Mankind, was a Carpenter in Nazareth and worked with His mortal father, Joseph. I believe that Christ was a very skilled and competent Carpenter. He spent the first three decades of His life assisting Joseph in the family carpenter shop"

Christa then stops her pace, ponders and says to David with a glint of pride,

"Since Christ also organized life on the Earth and other cosmic dwellings for life, He must be a very competent scientist and understands physics even better than my brilliant husband."

David responds,

"Indeed, Christ fully comprehends all truths in physics, and especially if M-Theory is valid and has merit."

Returning to their rental car, Christa asks David if they can take the tourist boat that crosses the Sea of Galilee from the eastern shore to Tiberius on the western shore and then return. David agrees, and they drive to the boat pier and board a tourist craft that regularly traverses the Sea. Christa tells David that Peter and Christ often sailed this sea.

While transiting the Sea of Galilee, the wind frolics with Christa's banded blonde tress that David treasurers. He lightheartedly warns Christa that if a violent storm arises while they are crossing the water, he cannot calm the water as Christ did with the Apostles aboard. However, Christa remarks that she knows David is an excellent swimmer and he will rescue her if a storm arises.

Fortunately, the Sea of Galilee is calm and the setting sun along the eastern shore casts a stunning sunset on this unique fresh water body in Israel. The couple returns to the eastern shore of the sea and Christa embraces David as she witnesses the serenity of nature's backdrop and thanks him for this treasured experience on the Sea of Galilee.

The couple then travels to Jerusalem, the Holy City for Jews, Christians, and Arabs. Jerusalem is the epicenter of the religious world for these three faiths that comprise over half of the world's population. As they travel the highway along the Eastern Wall of the Old City, the Dome of the Rock and Zion's Gate appear. Tears come to Christa's pale blue eyes and she moves to the curb and stops the rental car. David sees that Christa is emotionally transfigured as she gazes intently at the Eastern Temple Wall and Zion's Gate. Finally, Christa utters it is through that Gate that Christ will reenter when He returns to the Temple Mount and displays Himself to the astonished residents. David tells Christa to park the car and they will walk along the Temple Mount. The sun is now setting in the West and David points out the significant sites including the Dome of the Rock.

Wiping tears from her eyes with a tissue, she says to David,

"This is a lifelong dream for me that has come true. To be in the Holy Land where the Savior dwelt and traveled and presented His Gospel of 'Good News' for all mankind. And to be here with my beloved husband, a member of the House of David and a Disciple of Christ is an overwhelming blessing. David, I love you dearly, thank you for making me your wife and companion and bringing me to Israel."

The couple embraces and then travel to their hotel. They have reservations at the New Jerusalem Hotel near the Knesset, the Legislative Center for the Israeli Government. That evening they take a taxi from the hotel and travel to the Damascus Gate. Exiting the taxi, they enter the old city through the Damascus Gate. David escorts his bride along the crowded passageways filled with shops and visitors. Christa asks David to bargain at one of the shops for an Olive Wood Statue of Christ. David barters in fluent Arabic with the young boy Shop Keeper. At the close of a vigorous exchange, David declares that he can purchase the statue for 20 Shekels, a little less than 4 U.S. dollars. Christa tells David to buy the statue. While the boy is wrapping the statue with tissue paper, she gives the boy a U.S. dollar from her purse as a tip.

The couple walk to the plaza where tradition identifies the court where Roman Procreator, Pontius Pilate declared in Latin, "eco homo," or "behold the man" as Pilate displayed Christ to the Jews in the court. From there the couple then stroll the "via Delarosa." This Latin title denotes the "Way of Sorrows or Way of Suffering," that is marked by the suppositional Stations of the Cross along the path that Jesus dragged His cross. Tradition declares that Christ was driven to His crucifixion site at the Twelfth Station.

The next day, David and Christa travel to the BYU Center where David will be baptized and confirmed a member of the Church. At the BYU Jerusalem Center, the Stake President and two elders conduct the singular Baptism Service for David ben-Steinmann and later confirm him a member of the Church of Jesus Christ of Latter Day Saints and bestow the gift of the Holy Ghost. Afterwards, by special dispensation from the First Presidency of the Church, the Stake President and his counselors ordain David an Elder in the Church Jesus of Christ of Latter Day Saints.

Christa is filled with incomparable joy and righteous pride as she witnesses these Sacred Ordinances. David, her husband, is now a baptized member of the Church and holds the Melchizedek Priesthood. David's promise to Christa is fulfilled, and she voices a silent prayer of thanksgiving to her Heavenly Father as she witnesses these binding ordinances. Christa then recognizes to consummate this goal; she and David must be married and endowed in the Boston Temple after their return to the United States.

Christa entry in her Journal while in Jerusalem (15Jan)

After a fulfilling day of essential religious rites, David and Christa return to their apartment in Jerusalem. They quickly dress and then they slowly and casually stroll along the crowded streets of the Old City of Jerusalem as they seek a small, secluded Jewish café for their evening meal. They find a small café, along a side street, where they stop to eat. Again, David orders their dinner in Hebrew. Closing this very eventful day, they return to their apartment to close the day. After their individual and joint prayers before retiring, Christa collects her Journal from the nightstand and pens the following entry while David opens his Hebrew Bible and articulates a few verses in the language of the Jews to Christa's great delight.

Christa's weekly Personal Journal for 10-16 Jan 2011

I am now the wife of David ben-Steinmann and we are on our honeymoon. We are touring Jerusalem and David believes I am already pregnant. Incredible . . . he may be correct. We are already starting a family . . . an eternal family and my first child with David. I hope the child is a son for his father David.

The visit to the Holy Land has been an incredible experience for me. We have visited Tel Aviv, Caesarea, Haifa, Capernaum, the Sea of Galilee, Jerusalem, the Temple Mount, the Western Wall, and the BYU Jerusalem Center. Later we will tour Hebron, the Dead Sea, Masada and Beer Shiva.

I will have visited each of these marvelous sites accompanied by my husband, a son of David and a member of the Tribe of Judah.

Thank you Heavenly Father for this great blessing. David is now baptized, has received the Gift of the Holy Ghost, and holds the Melchizedek Priesthood. We will be married in the Boston Temple upon our return to America and my dreams and aspirations for us will then be realized.

CHAPTER 19 (17 Jan-23 Jan)

David and Christa's visit Mossad in Jerusalem (20Jan)

While David and Christa are resident in Jerusalem, David receives a telephone call in the morning from the Mossad, the Israeli Secret Service. The message is from a source that David knew at the Technion while David was a student there. The source is now with the Agency and wants to meet with both David and his new wife.

David informs Christa that after leaving the Technion, when his scientific career appeared very promising and later after receiving the Weissenhausern Endowment, David was approached by certain undercover members of the Jewish Embassy in Washington. They asked him if he would serve as a resource scientist and provide information regarding certain Middle East scientists and their military related projects with which David may become familiar. David acknowledged their request, but he has had very limited involvement with Mossad to the present time. David then quips to Christa.

"Perhaps it is my new bride that has provoked this renewed interest by the Israeli Mossad. You certainly would make a beautiful attractor for Arab and Muslim Scientists. A blond, blue eyed, statuesque female agent might prove invaluable and irresistible to Israel's enemies."

Christa frowns and rebukes David's assertion.

"David, I am consecrated to one man whom I love unconditionally and absolutely. My allurements, whatever they may be, are not displayed or available for anyone other than my handsome Jewish husband who is now my eternal companion and a Disciple of Christ."

David squeezes Christa's hand gently and says,

"Well spoken, my love."

David and Christa take a taxi to a small, unremarkable three-story hotel near the Israeli Knesset. They walk the heavily worn stairs to the third floor and are greeted by two men as they arrive on the threshold of the third floor from the narrow stair well. The men quickly greet David and Christa and usher them into the small apartment. After the apartment doors are secured, the greeters exchange informalities inquiring about David and Christa's visit to Jerusalem. The two men congratulate the couple on their recent marriage and compliment David on his stunning choice of a gentile bride. David agrees with their assessment of Christa and he states that Christa was his nurse at the Hospital where David was admitted after his serious automobile accident. It is due to Christa's due diligence and medical competence as his caregiver that his is now walking. The older man with a small, embroidered kipa on his baldhead and a short grey beard then provides Christa with a background regarding the Mossad.

"The Mossad was formed in April 1951 by David Ben-Gurion, the first Prime Minister of the State of Israel. The Mossad was and is essential in securing and preserving the State of Israel in a hostile, post-Second World War world."

Description of the Israeli Mossad

"Mossad has eight departments, but details of the internal organization are classified and cannot be divulged. The departments, which we want David to support, are the Research Department and the Technology Department. The Research Department is responsible for intelligence production, including daily situation reports, weekly summaries and detailed monthly reports. The Department is organized into 15 geographically specialized sections or desks, including the USA, Canada and Western Europe, Latin America, the Former Soviet Union, China, Africa, the Maghreb (Morocco, Algeria, and Tunisia), Libya, Iraq, Jordan, Syria, Saudi Arabia, the United Arab Emirates and Iran. The Technology Department is also responsible for development of advanced technologies for support of Mossad operations."

The Mossad Agent then focuses his comments to Christa.

"A joint nuclear desk in the Research and Technology Department is focused on issues related to special military weapons. The Iranian efforts to produce highly enriched U-235 are of great concern to us since Iran may be seeking to develop small, tactical nuclear weapons. Because of David's Jewish background, his international reputation, and his access to scientists in the Middle East scientific community, we need his support and your understanding for David's assessments of these dangerous developments. The principal aspiration we have for you is your support and participation in social functions associated with relevant David's scientific meetings and activities."

The elder Mossad Agent then implores Christa with the plea,

"Mrs. Steinmann, we need your support and cooperation as your husband interacts in these scientific conferences and meetings especially with scientists that are associated with the Iranian nuclear program. We sincerely hope that you will support us. Israel is a small, fragile nation and the only elected democracy in the Middle East. I know that you are a devout Mormon and I believe you support the existence and role that the House of David will have in the Second Coming of your Messiah. I read in your religious literature that you expect Christ will appear in the City of David when your Messiah returns."

Christa affirms his statement and declares that Christ was a Jew and a very devout member of the Tribe of Judah. She also reaffirms her love and full support for David, his posterity and the Jewish people. She then adds,

"David and I believe that Christ loves His Chosen People, the Jews, and Christ will be recognized and received as the long sought Messiah when He returns to the city of Jerusalem in the Last Days and declares His identity and Messiahship."

David smiles at his bride's strong testimony and declares to the Mossad agents,

"Gentlemen, you see I have married a very devout Christian as well as beautiful young women and companion. Furthermore, I fully endorse her words and embrace her Christian religion that is also now mine."

The two Israeli Mossad agents acknowledge David's words and then they affirm to Christa.

"Please understand that your husband's assistance and hopefully your support of our work are fully consistent with and known by the United States Intelligence Community. The security of your country is essential to the security of Israel. Moreover, your participation would be of great benefit to the tranquility and security for both of our countries and our people. The health and preservation of democracy is indispensable for peace throughout the world and particularly in the Middle East. Mrs. Steinmann, obviously you have great influence over your husband, Dr. Steinmann. I am aware that he was baptized into your faith, the Church of Jesus Christ of Latter Day Saints, during his visit here in Jerusalem. We are sorry to lose David to your Christian Faith, but we respect and honor Mormonism as a very good friend and supporter of Israel."

David and Christa then receive specific suggestions and instructions regarding their recommended responsibilities and the needs of Israel to protect the country and its people. After extended exchange of information, the joint meeting with the Mossad closes and David and Christa return to their apartment in Jerusalem.

Mossad meeting ends (23 Jan)

Upon returning to their Jerusalem Hotel, David and Christa discuss at length their meeting with the Mossad. David assures Christa that her only role in their service for the Mossad will be her role as David's wife and very gracious companion at any scientific or technical events that David and Christa attend. David has witnessed Christa's social skills and her cordial demeanor and tactful sociality in any social venue. He is fully confident that the requests by the Mossad will not be difficult for Christa to satisfy.

Christa responds that her father was a US Naval Pilot and held a Top Secret Security Clearance required for many of his military missions. She is familiar with the security and intelligence needs of both the US and Israel. She will honor her husband's support of Israel that now commands her allegiance and support as well.

The couple then eats at the Dining Room at the Hotel, and David and Christa review the remaining trips and events on their agenda for their visit to Israel. Tomorrow they will travel to Masada, the ancient mountain fortress built by Herod the Great and the location of the last stand by the Essenes when they defied the Romans in 70 AD. Masada is a national Israeli Historical site that demonstrated defiance against external enemies that sought to destroy Israel and crush the Jewish people. David informs Christa that it is required of each member of the Israeli Defense Force or IDF to visit Masada and learn of the heroism of its Jewish defenders against the Roman Legion sent to destroy them. David declares that it is this national resolution that preserves the Jewish state today. Christa smiles at her Jewish husband and replies,

"David, I am also a beneficiary of the Abrahamic blessings bestowed upon the Tribe of Judah and defender of the Jewish state along with my beloved husband. As Ruth declared in Ruth 1:16 in the Holy Bible: 'Where you go, I will go, and where you stay, I will stay. Your people will be my people.' "

Christa entry in her Journal while in Jerusalem (23 Jan)

The couple visits Masada and remaining sights that David has planned for visits by his bride. The last evening of their visit to the Holy Land before their return to the Boston, the couple enjoys an intimate and private evening discussing the exciting scenes that Christa has witnessed in the Tel Aviv, Haifa, the Sea of Galilee, Bethlehem, Hebron, and finally the city of Jerusalem and the Temple Mount in the Old City. Christa is emotional as she declares her appreciation for this honeymoon, and a visit to Israel and the Holy Land with her companion she loves so dearly. After their individual prayers, Christa asks David if he will be the voice for their joint evening prayer because she is overcome with gratitude to be David's wife and to be here in Jerusalem. If she attempts to vocalize their joint prayer, she might break into tears because of her feelings. David offers up their joint prayer as Christa holds David's hand.

David holds Christa in his arms upon retiring and Christa collects her Journal from the nightstand and tenderly pens the following entry:

Christa's weekly Personal Journal for 17-23 Jan 2011

We are now in Jerusalem. David has been baptized and ordained by Elders at the BYU Center and now holds the Melchizedek priesthood. My joy for these blessings is humbling and my gratitude to my Heavenly Father for David's membership and Priesthood is immeasurable. David has promised we will enter the Temple to be sealed to together when we return to the US.

We had an interesting visit with representatives of the Israeli government, but I will not record the details of that meeting in this journal because of the nature of the issues disclosed during the meeting.

Remarkably, I am now a member of the Steinmann Jewish household...and I am the proud matriarch in this House of Israel, even though David tells me that the claim to be Jewish is associated with the lineage of the mother of the house. Regardless, I am proud of my husband's ancestry, which I honor.

CHAPTER 20 (March 2011)

David and Christa return from Israel to David's apartment (March)

Two months have passed since their marriage by Christa's LDS bishop followed by a remarkable and inspiring visit to Israel and a honeymoon filled with passion and intimacy. David and Christa have now settled into David's apartment in Boston. It is quiet spring Sunday afternoon. David is in his study following church, while Christa is reviewing their wedding pictures and memorabilia that rest on the living room table. Wistfully, Christa ponders over her life with David as his wife. The marriage service was conducted in the Hospital's Administrator's Reception Room at Massachusetts General Hospital. Gathered in the Room for the marriage were many friends and acquaintances of David and Christa. Christa views the pictures taken during the brief ceremony and identifies the guests. She recognizes the President of Harvard, the Dean of the Medical School and of many of the staff at Massachusetts General Hospital who were acquainted with and befriended David and Christa. The actual ceremony was short, less than 10 minutes; but their union and warm greetings and sincere congratulation made the occasion what Christa had hoped and expected. These past two months as husband and wife have been filled with activities, travel, intimacy, and immense joy for the David Steinmann household. Christa's life has been profoundly transformed since she met David and she is now his wife and eternal companion.

As Christa returns the treasured wedding pictures to their proper location on the table, she ponders the remarkable changes wrought in her life since her first meeting with David in a Hospital Suite at Massachusetts General Hospital.

David and Christa conceive a child. (Mar in Boston)

One evening, as they are kneeling together in the bedroom of their Belmont apartment to offer their individual and joint prayers, upon closing their prayer together, Christa, still on her knees, turns to David and says,

"David, I have missed my menstrual period for the second consecutive month. Does that mean I am pregnant?"

David smiles at his beautiful bride and with firm confidence declares,

"Yes, sweetheart, my prominent Jewish proboscis informed me you were with child. I believe we conceived our first child on Thursday, 16 January 2010 while we were travelling to Jerusalem."

Christa is overwhelmed with maternal joy and she eagerly embraces David declaring,

"Oh David, I do hope you are correct. Oh, I know that you are correct. Your sensory perception is infallible, Dr. Steinmann."

Christa, still kneeling, sighs and leans back resting on her palms. She smiles with an angelic, maternal façade and expresses,

"Our first child…a son perhaps. We shall name him David, after his father."

Again, they passionately embrace. Then David smiles and comments pensively to Christa,

"David 'ben-squared' Steinmann. "Ben-squared is the mathematical expression for David ben-ben-Steinmann. However, we will not encumber him with that Hebrew title and I do not approve of David Steinmann, Junior or David Steinmann the Second. Those sobriquets are anathema to Jewish families."

David then says with a smile.

"Perhaps, David Olsen Steinmann? We can drop the ancient Jewish tradition of ben for 'son of' Steinmann."

Christa, radiant in David's eyes as a new Jewish matriarch, then interjects hesitantly,

"David, it could be a daughter…would you be disappointed?"

David looks upon his beautiful young wife, his 'Eve' in their Garden of Eden and says,

"Christa, I would treasure a daughter by you. I would witness you as a child… a beautiful, blond, blue-eyed charmer that I could spoil and cherish as her devoted father. To have two time displaced versions of you in my life would be an extraordinary blessing in the Steinmann family…you as a child and also as a mother and my eternal companion."

David then pauses and interposes,

"Perhaps I could persuade our daughter to become a physicist. A theoretical physicist…perchance a female, Nobel Laureate physicist. Did you know that only a few women have been awarded the Nobel Prize in Physics? Madame Curie received the prize in physics for her discoveries in radioactivity in 1903 and later, a Nobel Prize in chemistry for the discovery of radium and polonium in 1911. Remarkably, women have made great contributions in science and physics, but selfish, chauvinistic male Nobel judges have often deprived them of these awards. Examples of this deceptive chicanery include Lisa Meitner for recognizing nuclear fission in 1938 and Rosalind Franklin for experimentally disclosing the double helix DNA genome with her superb x-ray photographs."

Christa redirects David's response to the germane question,

"So, David, you will be satisfied if our first child is a boy or a girl?" David confirms affectionately,

"Any child by you would be a blessing and an honor in the Steinmann family. Christa, in traditional Jewish domestic relationships, it is the mother of the house who dictates the family's heritage and traditions. It is even possible my mother will accept my marriage to a Christian gentile when she witnesses her infant granddaughter or grandson. The love of a Jewish grandmother for her grandchild can surmount any barrier, even a lifetime of sustained bigotry and intolerance."

Christa then hesitantly objects,

"But we will not name our daughter, 'Sariah,' will we?" David responds quickly and emphatically,

"No."

David then responds as an emotional afterthought,

"I will not burden our daughter with that onerous epaulet. That would awaken deep-seated, painful memories in my own past."

David resumes teaching and research at Harvard

David has resumed his regular teaching and research duties at Harvard University. His visit to the LHC at CERN has provided him with new directions and incentives for his research in M-Theory and gained renewed experimental support to support his hypothesis for dimensional confirmation. The major challenge remains to demonstrate experimentally, the existence of additional spatial dimensions beyond the four dimensions of our known spacetime universe. If he can establish experimentally, the transient existence of at least one additional spatial dimension during the violent interaction of proton collisions in the beam aperture of the LHC, then M-Theory will assume a credible and unique role in the archives of cosmology.

David has acquired two additional postdoctoral students to support his earlier group he had at the Hospital when he conducted the seminar classes in his Hospital Suite. He believes with this new infusion of student help and the incorporation of octonions to address the transition of the very rapid decay of collision products from the LHC particle burst at CERN will be successful. He can direct the experimentalists as to what to seek as they analyze the complex array of collision products and follow their ephemeral changes in energy state.

David's hypothesis is that during these very short time intervals when the Terra-eV particles interact and collide, that conditions similar to those during the big bang, may exhibit, partially or fully, the multidimensional conditions that are postulated in M-Theory. Assumedly, particles in that brief, transitional condition will distribute their energy and mass throughout more dimensions than our present spacetime conditions exhibit. These remnants of energy and mass may now be manifest in the dark energy and dark mass now evident in our universe. David describes these conditions to Christa and uses her innate ability to grasp and assess their rationality as a credibility check for his cosmological conjectures. Remarkably, Christa is very insightful as usual and challenges David when his reflections and conjectures appear to be unwarranted or illogical. David often explores and attempts to assert his ideas to Christa for her rationality check. To David's delight, she has often motivated him to explore a different scheme, as she demands David's basis and reasoning for his ideas.

David often holds his graduate seminars in his home with several of his students including Christa as a participant in the discussions. Of course, she also serves as the hostess and provides the seminar participants with fresh fruit, punch, and homemade cookies. Occasionally, David's students even bring their small children with them to the seminars; and it is Christa's responsibility and delight to entertain these children during the seminar. These seminar participants often confide to David, it is his wife's hospitality and graciousness that makes his seminars at his residence a pleasure to attend.

David and Christa are married and sealed in the Boston Temple

In a sacred and spiritual ceremony, Christa's culminating desire for her and David's marriage is their sealing as an eternal couple performed in the Boston Temple. This event transpired on a beautiful spring day in late March as David and Christa Steinmann were sealed together eternally. Friends of the couple gathered in the Steinmann home the following evening to congratulate them and become better acquainted with Christa and her husband and companion. After a joyous but exhausting evening and the departure of their guests, they prepare to retire for the evening. Following their closing prayers with David as the voice for the kneeling couple and ardent embraces, Christa removes her journal from her nightstand and enters the following entry while David reviews a recent scientific paper.

Christa's weekly Personal Journal for March 2011

We are now in Boston and David has resumed his teaching and research at Harvard. With great pleasure and humility, dear journal, I proudly announce that David and I were sealed as a couple in the Boston Temple and we are now eternal companions. Oh, my joy is overwhelming. My gratitude to my Heavenly Father for David's membership in the Church, his Priesthood, and now the Temple sealing of our marriage is immeasurable and humbling.

However, my greatest joy at this moment is my possible pregnancy. I may now be the host of David's child within my womb. If this is confirmed by my pending visit to my obstetrician, I will become a mother in Zion. What a marvelous blessing, our Heavenly Father has bestowed upon his matriarchal handmaidens to be the fountain of His heavenly spirits born into mortality. To serve as the instrument by which children come to earth for mortal testing and salvation is a blessing for me. I now share some of the joy witnessed by Christ's mother Mary when she was informed by the Angel Gabriel, that she would be the mortal recipient and vessel for the Father's only begotten Son in the Flesh, Jesus Christ.

CHAPTER 21

Christa's pregnancy confirmed

Christa visited her obstetrician this morning and her female physician confirmed what David had told Christa earlier. She was pregnant! She had conceived during the very first week of their marriage as David had declared would occur with complete confidence. His sensatory gifts were remarkable and still inexplicable to Christa. Their first child had now been conceived, and David and she were overjoyed.

Christa silently recounted her life altering events over the past exhilarating six-month period. A tragic accident in early October of 2010 had taken the life of David's sister Rebecca and almost ended David's life. She had witnessed both David and Rebecca's admission and their traumatic treatment in the Emergency Ward. Rebecca's injuries were too extensive, and she succumbed to the accident. David's injuries were also life threatening, but God's intervention and the skilled and effective emergency staff at Massachusetts General Hospital saved his life. She was so grateful that the trauma emergency unit had acted quickly and successfully, and now David was alive, ambulatory and her husband...the Father of the child she carried. She then pondered the results of this tragedy. Even though the accident was a tragic event for David and his sister Rebecca, the accident had brought David into her life and she and David were now husband and wife with a child developing within her womb.

Christa recognized God's intervention even in this dreadful accident and how great blessings can emerge from human adversity and result in abundant blessings for her life with David. She loved David with passion, respect and honor. Smiling, she internalized her new status...Christa Steinmann, RN, was now the wife of one of the eminent young scientists of this age. She was certain that David would garner the Nobel Prize and confirm that M-Theory was the final union...or perhaps...'the celestial marriage' of the physical sciences.

David had joined the Church and been baptized and united with Christa in a Celestial, Eternal Marriage Ceremony in the Boston Temple. Christa had always planned for and expected a Temple marriage since her childhood. But to have won the hand of a handsome, famous Jewish scientist and former agnostic and brought him into the Gospel of Jesus Christ was an incredible blessing and honor. She was an effective missionary even without a formal calling by the Church. Christa mused tacitly,

"God certainly works in marvelous ways to perform his wonders. Out of tragedy have come great blessings."

Christa could conceive of no other blessings that the Lord could bestow upon her that were not hers now or would soon be hers. Her heart welled upon in gratitude and praise, and she quietly voiced,

"Thank you Father in Heaven, I am truly blessed and honored with Thy Hand in my life."

She then corrected her words, murmuring, "With Thy Hand in our lives...David's and mine, together, forever."

Later, after their joint prayers and retiring, Christa continues to ponder these events as she lies by David's side in their King Size Bed in their Belmont apartment. Massachusetts. David has promised her that in the summer they will purchase a home...a home for them and their first child. She turns her head towards David who is sleeping soundly by her side. Christa smiles as she realizes that David's warm hand is gently resting on her abdomen. She is so grateful for him and his love for her. His embracing hand acknowledges in her mind that David is unconsciously safeguarding his child, as a righteous father should. What does their new life together hold for them and David's career in physics, and now a more important career as a husband, a father and her...her Eternal companion.

David's Grand Mal seizure (Apr)

However, Christa does not realize that this night of gratitude and bliss will end with an extraordinary event that will alter her life again. This time, however, the event will bring unspeakable grief and great tragedy for her. Moreover, her mortal life will again be dramatically altered.

While Christa wistfully ponders her blessings, David's hand, suddenly twists and writhes as he quickly removes his hand from her abdomen. She then hears a deep, guttural sound from David's mouth. Soon his entire body is jerking and violently shaking the bed. She realizes David is suffering an epileptic seizure…a very intense seizure. She has feared that he might experience such a seizure, especially now that he is not taking the previous regimen of drugs that have controlled his epileptic seizures in the past. She quickly rises from their bed, stands, and goes to David's side of the bed and turns on the lamp on the nightstand. With the bright light illuminating his face, she sees he is gagging and appears to be gasping for breath. She quickly grabs and pulls a handful of tissue from the nightstand and forces the wadded ball to the side of mouth between his teeth so that he will not choke on his uncontrolled tongue sweeping through his mouth. She watches David's seizures with fear. Will his seizure transition into a dreadful 'Grand Mal' seizure?

Christa strips the bed covers from David and again quickly checks his breathing. His respiration is fast and shallow, bordering on hyperventilation, and his spasms shake his body in a random cadence jolting the bed. Christa remembers that she has a prescribed emergency dosage of potent Gabapentin that David has previously used in the past to suppress severe epileptic seizures. But that was before his automobile accident and the major serial surgeries he underwent to restore the function of his right leg. The potent drug is in the cabinet in the bathroom. However, administering this narcotic may result in unknown complications because of David's multiple surgeries and compromised immune system. The damage to his right hip and femur has severely limited his erythrocyte production and affected his immune system. However, this seizure is a dangerous development and she weighs the risks of administering the dose or waiting for a termination of the seizure. She decides she must seek emergency help.

Christa calls 911

She goes to the telephone on the nightstand and dials 911. The emergency operator answers and Christa declares a medical emergency.

"A 31 year old male patient with confirmed epilepsy is presently experiencing a serious seizure. This seizure may escalate into a violent 'Grand Mal' stage and for this patient and such a state could be fatal. I am a registered nurse at Massachusetts General Hospital and the patient is my husband. Immediately dispatch an emergency medical team to 1020 East Worcester in Belmont, Massachusetts, apartment number ONE on the ground floor. Please repeat my verbal request for confirmation of my emergency request."

The dispatcher repeats the emergency message to Christa responding that the emergency unit will be dispatched to 1020 East Winchester in Belmont, Massachusetts, apartment number ONE, ground floor. The dispatcher then states that the only available emergency unit in the area has responded to a nearby fatal auto accident, but will be available afterward for dispatch to her location.

Christa responds anxiously,

"Please accelerate their response; tell them the victim is in an escalating stage of epileptic seizure and is now hyperventilating."

As Christa closes the emergency telephone call, she attempts to gently constrain David's spasms and insure his respiration is viable. David's legs and arms are now flailing violently. His arm strikes the nightstand overturning the night lamp and the telephone that spill across the bedroom floor.

Christa realizes that a grand mal seizure, also known as a generalized colonic seizure, is the most serious type, one that involves the entire body and typically causes the victim to lose consciousness and even death.

Christa administers medical dose

David's seizure is rapidly worsening, and Christa now concludes that she must administer the Gabapentin immediately, even with its possible side effects. David has lapsed into a fully acute, aggravated epileptic seizure stage and is exhibiting severe convulsions. She steps around the debris on the floor and rushes to the bathroom. She opens the cabinet, removes a 5 cc syringe, and quickly inserts a 22 gage, subdural needle onto the neck of the syringe. She grasps the unopened vial of Gabapentin, removes the aluminum cap, carefully mixes the contents to avoid air entrapment, inverts the vial and pushes the needle of the syringe into the opaque fluid. She withdraws a maximum dose, 1.5 cc or 1500mg, from the vial that reads, "Administer only for suppression and control of severe epileptic seizures".

Returning to David's bedside, his Grand Mal seizure is now fully evident. Christa attempts to grasp David's left arm that is violently thrashing with the rest his body. To constrain his movements so she can deliver this injection, she sits on David's legs, pushes his left arm to the bed and rapidly sinks the needle into his upper arm.

The quelling effect of the massive dose of Gabapentin is rapid. David's body soon relaxes, and his uncontrolled movements slowly subside. As Christa sits on the edge of bed adjacent to David, she is relieved and believes the severe seizures are now under control. She anticipates the possible chronic suppressive effects of the drug on his immune system can be corrected later when the emergency crew arrives or at the Emergency Ward at the hospital.

Christa gazes upon her husband, as he lies motionless on the bed. She is grateful for the drug's effective sedating effects and suppression of

David's seizure. His respiration has stabilized, and the spasms have now ended. She sees that David's relaxed face is covered with perspiration from the rapid muscular motion during the seizure. She recovers the tissue box from the floor and removes several tissues. While using these tissues to wipe David's face, she remembers David's previous seizure in the Hospital and her spontaneous act of love with a kiss to her unconscious patient. Christa smiles and kisses David's forehead with ardor. He is now her eternal companion, and she is free to display her love without restrain or criticism. There is no head nurse monitoring her actions now.

Christa's is grateful for her medical training and her ability to control quickly David's medical emergency. She muses that medical skills are an area in which she exceeds her husband's many talents. She is grateful that she can provide this valuable service for David. When the emergency unit arrives, she will inform them of her acute administration of 1500 mg of Gabapentin, and the EMTs can monitor his physical signs for correction of any after affects associated with the Gabapentin. She has successfully terminated David's Grand Mal seizure, and he is resting quietly now. However, she is anxious for the arrival of the emergency team and their additional oversight and support.

David's unexpected response to Christa's medical intervention

While Christa is pondering over David's seizure, David's body suddenly wells up, convulses, and then slumps back onto the disheveled bed. His arms fall lifeless to his side. His head turns, and his eyes widen and become fixed and dilated. Christa is alarmed and frightened at this severe, acute response to the Gabapentin. Was the dose too large and its effects to severe? Is David experiencing a prompt, anaphylactic shock from the depressant impact of the drug?

She quickly removes the wadded tissue from his mouth. As she clears small tissue pieces that he bit upon during the seizure, Christa suddenly realizes that David is not breathing now. She instinctively checks for a pulse in his right carotid artery. She can feel no pulse in the right artery. She turns his head and presses the left carotid artery; still no pulse. David is in full cardio-pulmonary arrest. Instinctively she presses her index finger into his mouth moving his tongue aside. She encloses her mouth over his and forces air into his lungs. His chest rises in response to her intubation. She then presses the heel of her right hand backed by pressure from her left hand to deliver a strong compression of David's diaphragm below the sternum. She anxiously counts as she serially compresses his chest. One…Two…Three…Four. She then turns and covers his mouth with hers and again drives air into his lungs. She is well trained in administering emergency cardio-pulmonary resuscitation, and she now responds instinctively and reflexively from her years of training in such emergency medical conditions. She continues these cardio-pulmonary (CPR) cycles watching for David to regain breathing and cardiac response. Again, she quickly probes his carotid artery for a pulse…still there is no discernible pulse. He is still not breathing nor; does he have an arterial pulse. Christa is now deeply concerned and fearful, but she realizes she must continue her emergency procedures and control her fear and anxiety over David's critical state. Her experience with CPR in emergency practice has conditioned her to contain her anxiety and fear and administer lifesaving procedures properly and effectively. As she continues her exhausting efforts to resuscitate David she expects that the emergency ambulance with full CPR equipment should soon arrive soon…why don't they come? However, she must continue administering CPR on David's unresponsive body.

David's NDE and Christa's resuscitation attempts (Mar)

As this drama of frantic lifesaving efforts by Christa ensues, David is now separated from his body and is observing Christa's desperate actions to resuscitate him. David is again in an out-of-body state or NDE, as he experienced earlier with his automobile accident. He observes his limp body, motionless on the bed and his lifeless eyes staring towards the ceiling of the bedroom. Although he feels no stress or pain, he witnesses Christa's growing terror as she desperately acts to preserve David's mortal life. He then hears Christa cry out in anguish.

"David! David! Breathe...regain your cardiac function...David, do not leave me!"

Christa continues her exhausting CPR procedure. Perspiration now covers her flushed face...tears stream down her pale cheeks as she methodically continues her resuscitation efforts on David. Only her extensive experience of nursing discipline and practice contain her panic and sustain her conditioned actions to save David's mortal life. Again, she vocalizes her anguish as she utters in despair,

"Oh God, dear God, help me save him. I must save him...he must live! David...David!"

David pleads to return to Christa

As David witnesses these frantic actions and desperate words of Christa, he becomes aware of the presence of another Being. The Being is unknown to him, but the Being's dominant and compelling influence commands David's attention and he recognizes the absolute power and authority of this Entity. The Being informs David, without words but with certainty, that David is now dead! He will not return to his mortal body as he did before when David and his sister were at this point of mortal transition following their automobile accident. His mortal life was preserved then for his mortal conversion and salvation, but these purposes have been accomplished now.

However, David, witnessing Christa's continuing and frantic actions to restore his mortality, counters the Being's message and asserts that he must return to his mortal body. Christa is desperately struggling and in great turmoil and fear as she attempts to resuscitate David. David dearly loves Christa and he must end her despair and return to support her and his child that she now carries. David fervently entreats the Being that Christa needs him and so does his unborn child.

However, the Being's unyielding response is that David must exit mortality and accompany the Being. The male fetus developing within Christa's womb will arrive safely into mortality and Christa will nurture and adequately provide for their male child. Christa will be preserved in mortality to sustain and rear the child to full maturity.

Again, David attempts to resist the Being's message and pleads that he be allowed to return to his mortal body and support Christa and his unborn son. David must not leave them now alone and unsupported by him. How can Christa support herself and the child without him? David must continue as a husband and mortal companion to Christa. David has the obligation to serve as the father to their unborn child.

Then David compellingly declares his belief in Christ and His Heavenly Father.

"My pursuit and findings in science have confirmed my faith in Christ. My testimony is full and certain. Jesus Christ is my personal Savior and my Advocate with the Father."

The Being responds to David that his faith in Christ is affirmed and his exaltation is secured. Christa and her child, whom Christa will name David after his mortal father, will complete their allotted time in mortality. Their son will be born under the covenant made in the Temple when David and Christa were sealed together. Their child will then be reunited with David and Christa at the close of the child's mortality. The Being informs David that Christa and the child will be his through the binding ceremony previously performed in the Temple sealing ordinance.

Again, the Being resolutely declares to David that his mortal work on the earth is concluded. There are others now awaiting him...his father...his sister, Rebecca, and his other family progenitors. These ancestors seek and require David's witness and testimony, so they can be sealed together as a single, eternal family. David has work to perform among the ancestors who await him beyond the mortal veil.

The Being then proclaims to David's mind,

"David's contributions to a fuller understanding of the true nature of God's Kingdoms in space and time have been completed. David's discoveries and disclosures in mortality are true and will accomplish the purposes that were prepared for his mortal sojourn. David's research and influence will produce great good and will expand and confirm the faith and hope of many scientists in the Gospel of Jesus Christ. But David's mortal tenure is now at an end."

The Being informs David that his mortal work is ended

The Being then conveys the final message to David.

"Mankind must come to Christ through hope and faith, not through scientific investigation and knowledge alone. The essential test for man in mortality is the trial of faith and belief in Christ and submission and obedience to His Commandments. Mankind is not meant or even permitted to have complete knowledge of God's Kingdom during this mortal probation; otherwise, there is no trial of mortal men's faith and works. Satan and his followers understand the Father's Plan and know of His kingdom, but they will not be participants in the Father's Plan because of their rebellion against the Father and opposition to His Son, Jesus Christ. The Father's Plan requires a test and confirmation of each earthly mortal's ability to seek the Father through faith in and obedience to His Son, Jesus Christ. This is the essential trial for mortals, to have confidence in salvation and the Atoning Sacrifice of Jesus Christ."

The Being closes message to David.

"To dwell with the Father in his Kingdom requires complete trust and abiding faith in His Son, Jesus Christ. Full knowledge and understanding of the true spatial and temporal dimensions belonging to the Kingdom of Heaven are not to be fully revealed or understood in mortality. This knowledge is reserved for those immortal, exalted beings that became Disciples of Christ, and therefore partake of the Father's glory and inherit a place in the Celestial Kingdom."

The Being closes the message to David with the final declaration.

"To dwell with the Father requires absolute faith and enduring confidence in Christ. Salvation and Exaltation for mortal beings are pursued and gained through faith, hope and charity."

David then departs from this mortal universe with the Being across the veil to join his ancestors and await judgment by Christ for David's deeds and desires while in mortality and later eternal reunion with his eternal companion, Christa and his son.

THE END

APPENDIX

The Higgs boson is an essential, Standard Model particle that has now been experimentally observed and confirmed at CERN. The existence of the Higgs boson is a consequence of the so-called Higgs Field, which is an essential the part of the Standard Model of Particle (SMP) and explains how the known elementary particles form matter and become massive. For example, the Higgs boson explains the difference between the massless photon, which mediates electromagnetism, and the massive W and Z bosons, which mediate the weak force. Since the Higgs boson has now been observed, it is an integral and necessary component of the Standard Model of Physics for the material world.

The SMP or "standard model" of particle physics is a system that attempts to describe the forces, components, and reactions of the basic particles (hadrons, leptons and bosons) that make up matter and energy. It not only deals with atoms and their components, but the pieces that compose some subatomic particles. Although this model does have some major gaps, including gravity, and some experimental contradictions; the standard model is still a very good method of understanding particle physics, and it continues to improve. The model predicts that there are certain elementary particles even smaller than protons and neutrons.

Each of the subatomic particles contributes to the forces that cause all matter interactions. One of the most important, but least understood, aspects of matter is mass. Science is not entirely sure why some particles seem massless, like photons, and others exhibit mass and are "massive." Now the standard model confirms that there is an elementary particle, the Higgs boson, which can produce the effect of mass. Confirmation of the Higgs boson has been a major milestone in our understanding of physics.

The "God particle" nickname actually arose when the book, *The God Particle: If the Universe Is the Answer, What Is the Question?* by Leon

Lederman. Since then, the God particle has taken on a life of its own, in part because of the monumental questions about matter that the God particle might be able to answer. The man who first proposed the Higgs boson's existence, Peter Higgs, is offended by the nickname, "God particle." Higgs is an avowed atheist.

The Higgs boson using the Large Hadron Collider, a particle accelerator in Switzerland, has confirmed the existence of the Higgs boson. As with any scientific discovery, God's amazing creation becomes more and more impressive as we learn more. The Higgs boson represents a step forward in human knowledge and our appreciation of God's universe. Whether or not the Higgs Boson is the "God particle," we quote this about Christ: "For by him all things were created: things in heaven and on earth, visible and invisible, all things were created by him and for him" (Colossians 1:16).

Astrophysics and Cosmology Theory are involved with theoretical particle astrophysics and big-bang cosmology as well as more speculative string theory inspired cosmologies. Understanding the quarks to cosmos connection is a recent focus of physics as well as better understanding the implications of the fluctuation spectra of the cosmic microwave background. The cosmology and astrophysics research supported is associated with people with training in particle theory and encompasses dark matter, dark energy, high-energy cosmic rays as well as exotic cosmologies arising from Brane-world and String Theory scenarios. This includes formulating new approaches for theoretical, computational, and experimental research that explore the fundamental laws of physics and the behavior of physical systems; formulating quantitative hypotheses; exploring and analyzing the implications of such hypotheses analytically and computationally and interpreting the results of experiments.

D&C 76 117 To whom he grants this privilege of seeing and knowing for themselves.

A warning to David in his exploration of God's Kingdom in Heaven is provided in the following religious scriptures.

> Moses 15, Wherefore, no man can behold all my works, except he behold all my glory: and no man can behold all my glory, and afterwards remain in the flesh on the earth.
>
> D&C 58 3, Ye cannot behold with your natural eyes for the present time the design of your God concerning those things, which shall come hereafter, and the glory, which shall follow after much tribulation

Albert Einstein said, "Whether you can observe a thing or not depends on the theory which you use. It is the theory which decides what can be observed."

David's final model of the universe.

1- The Big bang 13.8 billion years ago resulted in the formation of our present, observable Universe. With the occurrence of the Big Bang, our present three spatial dimensions and one time dimension came into being as a spacetime entity, with our universe, governed by General Relativity and Quantum Mechanics. Before the Big Bang, these dimensional and time metric did not exist in our present, observable Universe.

2- Our present universe has over a 100 billion galaxies each with hundreds of billions of stars and planets that we can physically observe. However, despite this large number of observable galaxies and stars and other celestial feature, we actually observe less than 5% of the total energy and matter that is required to explain the physical behavior of our universe based upon quantum mechanics and Einstein's spacetime mechanics. The remaining 95% that we cannot observe visually is referred to as dark energy and dark matter. We address these dark entities as "dark" or transparent because we cannot directly observe these features via visible light or the emission of electromagnetic radiation. We infer the existence of dark matter and energy to explain the accelerating expansion and the behavior and motion of our visible universe that we have observed and measured in our universe. These dark entities manifest their existence principally through gravity that affects both our visible universe and dark matter and dark energy components. The dilemma that confronts science is that observable universe has insufficient mass and energy to account for the behavior we observe. The existence of dark matter is essential for validity of the SMP, the observable composition and our understanding of our present universe. Correct identification of these dark entities is the major issue confronting cosmology,

3- Even though the "big bang" occurred about 13.8 billion years ago, our anthropological time and space did not then exist in our present universe, the duration of the existence for these dark entities (dark energy and dark matter) are eternal and existed before the birth of our own universe. To address the dilemma that space and time did not exist in our universe before the big bang, the creation of universes such as ours from primordial explosions is a common feature and is a continuing process throughout the entire multi-universe cosmos. We cannot observe, at least presently, these other multiverses since they are outside the spacetime domain in our visible universe. But, we conceive of these multiverses as growing bubbles, each expanding under the influence of dark energy and dark matter.

4- Our universe continues as a growing bubble of space and matter that is held together by both dark matter but is accelerating in size under the influence of dark energy.

5- The formation of our universe was not a singular, unique, or isolated event. New universes are being birthed, producing other universes that comprise a grand ensemble of multiverses. However, each of these multiverses is infused by the same dark matter and energy field that is the universal substance supporting all the multiverses. It is this dark energy and dark matter field in which God dwells and interacts with the multiverses.

These dark entities are invisible and absent the electromagnetic signature that exists in our present universe. Other universes come into being occupying their own position in spacetime. These universes undergo similar histories that our universe experiences, although their properties for matter and energy and behavior may differ from ours.

6- The theory for the generation and behavior of these multiverses is provided by String Theory or M-Theory. Dimensionality in these multiverses may possess 10 spatial dimensions and one time dimension that is bi-directional and can span forward in time and reverse in time. The prevailing physics and properties may differ within these individual multiverses. However, interactions between these multiverses are possible through the dark matter and dark energy that contains all these multiverses.

7- Matter and energy are distributed within each multiverse to a total unique value for the occupancy of all multiverses. For our universe the percentage of ordinary visible mater and energy is about 5%.

8- God(s) is able to access and observe each of the multiverses that exist throughout the grand ensemble that embraces all these individual multiverses spread though out space and time. God may utilize these multiverses and populate them with intelligences for the expansion of His work, which is to "Bring to pass immortality and eternal life of man."

David's Hypothesis Conclusion
The principal residence for the kingdom of God is within dark matter. This dark or transparent matter domain incorporates time and space entities that are multidimensional and are without the limits and constraints that we observe within our observable universe. The full behavior of matter and energy within dark matter is not governed or limited by the speed of light or the unidirectional flow of time present for our observable universe. These enhanced properties are determined by the conditions that exist within singularities or special conditions that can occur for matter and time.

Male Names

Aaron (Aharon)	אַהֲרֹן
Adam	אָדָם
Benjamin	בִּנְיָמִין
Daniel	דָּנִיֵּאל
David	דָּוִד
Jonathan (Y'honatan)	יְהוֹנָתָן
Joseph	יוֹסֵף
Joshua (Y'hoshua)	יְהוֹשֻׁעַ
Michael	מִיכָאֵל
Samuel (Sh'mu'el)	שְׁמוּאֵל

Female Names

Deborah	דְּבוֹרָה
Elizabeth (Elisheva)	אֱלִישֶׁבַע
Judith (Y'hudit)	יְהוּדִית
Mary (Miriam)	מִרְיָם
Rachel	רָחֵל
Rebecca (Rivka)	רִבְקָה
Ruth	רוּת
Sariah	שָׂרָה
Sharon (Bible place name)	שָׁרוֹן
Susan (Shoshanah)	שׁוֹשַׁנָּה

PROLOGUE

Background and Setting for "The Strings of God"

David ben-Steinmann - Theoretical Physicist

David ben-Steinmann is a brilliant, young theoretical physicist. Not yet thirty years old, he has secured a position of eminence in the scientific world of nuclear physics and in his specialty field of astrophysics and cosmology. Astrophysics is the scientific study of the behavior and history of stars – there birth, lifetime and death. Cosmology deals with the origin, history, composition and behavior of all entities within the universe or cosmos including the stars, planets, and other physical and energy entities within the known universe. Modern cosmology now contemplates the possible existence of parallel universes, which surround and may include our observable universes. The laws of physics as we observe them in our universe may be markedly different in these parallel universes.

David is Jewish as are many of his scientific contemporaries throughout physics. Furthermore, Jewish scientists have dominated this exclusive field of astrophysics and cosmology that seeks to disclose the origin, nature, behavior, and fate of our universe. David's scientific career and genius, however, is extraordinary and exceeds the other contemporary, young prodigies, both Jewish and Gentile, in his chosen field of science.

David ben-Steinmann or David Steinmann, as he is better known and prefers to be cited in his seminal and copious publications, is the preeminent, singular figure in the evolving field of physics called String Theory or Super String Theory. David has resolved many of the early errors and weaknesses surrounding String Theory and established its promising successor now known as "M-Theory." The single letter "M" is believed to be the contraction that denotes "membrane" for M-Theory. However, the actual source of the term is controversial and not well documented. Speculation is that the noun "membrane" alludes to the illusive boundary that possibly surrounds and isolates our known universe from other possible universes, or more descriptively, multiverses or parallel universes that M-Theory asserts probably exist. These parallel universes are postulated to resolve current conundrums in cosmology since much of the projected energy and mass required to explain the present behavior of the visible universe is presently unobservable. This energy and matter has been classified as dark energy and dark mass because they are outside the metrics or measurements of present cosmology. Recent, measurable enigmas such as the confirmed accelerating expansion of the universe and the anomalous behavior of the gravitational field of galaxies have resulted in the wide acceptance of these so-called "dark entities."

Merger of the very small (Quantum Mechanics) and large (General Relativity)

Scientists in the physical science recognize Quantum Mechanics or QM as the study of the very small structures in the universe. These structures include atoms and radiation or light and the other wide array of atomic and subatomic particles found in the "Standard Model of Particle Physics" with the acronym SMPP. These small physical entities bear foreign and capricious names such as alpha, beta, gamma, tau, mu, upsilon and lambda in Greek; and charm, quark, top and bottom in English. These entities have all been experimentally observed and confirmed. The recent discovery of the "Higgs Boson" that confirms the existence of mass in our universe and resulted in the Nobel Prize being awarded Peter Higgs who postulated this class of bosons and those experimentalists at the Large Hadron Collider who measured and confirmed the existence of this essential class of bosons.

On the other hand, the large material entities, more familiar to humans and visibly observable, include the infrastructures of plants and animals, Homo sapiens or humans, the earth, the solar system, the stars and even galaxies. These massive structures have been effectively and successfully described by Isaac Newton and later, more accurately described by Albert Einstein through his Theory of General Relativity and its corollary of spacetime.

Although these two major disciplines, viz., quantum mechanics and Einstein's General Relativity have achieved great success in accurately and comprehensively describing their respective domains in science, the two disciplines are incompatible and incommensurate when any attempt is made to merge them into a single unified theory for all physical behavior. Einstein spent the last years of his life attempting, unsuccessfully, to formulate a "Unified Field Theory" to merge the conflicting theories for the cosmos.

The great difference in size and scope of QM and relativistic spacetime theory associated with atoms and galaxies has obscured the irreconcilable barrier between them. However, this inconsistency is now fully evident and problematic with the discovery and observation of "Black Holes" where matter is compressed into an infinitesimal volume with gravitational fields approaching infinity. Furthermore, black holes are now appearing to exist within every galaxy of our universe and are essential features of galactic behavior. It is fully accepted in the physics community that the correct description of black holes now demands the merger of both quantum mechanics and General Relativity to provide a quantitative bridge. These conditions are unacceptable to the present community of world physicists. This is a colossal dilemma for General Relativity and Quantum Mechanics and the Standard Model of Particle Physics (SMPP).

M-theory is a direct attempt to address this dilemma. M-Theory is the controversial, audacious, but unconfirmed bridge using mathematical modeling and physical extrapolation that may provide a marriage of these disparate partners finally unifying the entire family of physics. M-theory asserts that to merge quantum mechanics (QM) and General Relativity's "four dimensional spacetime" requires a new paradigm with 11 dimensions and the elimination of point properties for all material particles to avoid the ubiquitous problem arising from singularities. Singularities appear in the

Standard Model for particle physics, when variables describing size and other properties approach zero and result in meaningless, infinite values for these variables and their associated parameters.

Planck Units in M-Theory

However, in M-Theory, all matter and energy, indeed everything, is composed only of open and closed vibrating strings of energy. The frequency and vibrational modes of these strings dictate the energy and mass for all cosmic entities and all their intrinsic properties. These open and closed strings are infinitesimally small with Planck size dimensions and wave characteristics. The strings continually undulate at complex frequencies providing the diverse macroscopic properties of mass and energy that we observe. M-Theory requires more dimensions, greater than four, to achieve this merger. Furthermore, Planck units provide the essentials limits for these multi-dimensions that preclude singularities found in SMPP. Also, associated with M-Theory is the possibility that alternative universes, often referred to as parallel universes exist. These additional universes may have their own unique characteristics and behavior. The laws of physics and consequent appearance and behavior of all entities found in these alternative universes may be very different from the entities in our universe. Also of interest, is the possibility that the big bang that created our present universe 13.8 billion years ago may be a common, repeating occurrence and there exists a multiverse comprised of innumerable universes typical of our present universe created by repetitive creation and expiration events.

Returning to M-Theory, it is assumed that all entities of energy and matter in our universe are composed of these Planck units and are appropriately small or large as necessary to complete the theory. Planck units, unlike the entities in the Standard Model of Particle Physics, are always finite and are assumed to satisfy the point properties for all dimensions encountered with the Standard Model of Particle Physics. A Table of approximate Planck entities is shown in the following Table.

TABLE I - PLANCK UNITS ASSOCIATED WITH M-THEORY

Planck Unit	Magnitude	SI Units
Planck Length	$= 1.6 \times 10^{-35}$	meter
Planck Area	$= 2.1 \times 10^{-70}$	square meter
Planck Volume	$= 4.2 \times 10^{-105}$	cubic meter
Planck Mass	$= 2.2 \times 10^{-8}$	kilogram
Planck Mass	$= 0.22$	micrograms
Planck Density	$= 5.2 \times 10^{96}$	kilogram/cubic meter
Planck Energy	$= 2.0 \times 10^{9}$	Joule
Planck Temperature	$= 1.4 \times 10^{32}$	Kelvin
Planck Time	$= 5.4 \times 10^{-44}$	second
Planck Momentum	$= 1.0 \times 10^{-34}$	kilogram-meter/second
Planck Electric Charge	$= 1.9 \times 10^{-18}$	Coulomb

Assessing these Planck units, the following properties are evident:

The size and time scales of string objects are extremely small, much smaller than the fundamental material particles that are encountered in the Standard Model for Particle Physics (SMPP). What is fundamental with Planck units is that although they extremely small, these values never go to zero or vanish in string theory, so singularities never occur as they often do in the SMPP.

On the other hand, the density, energy and temperature units (kg/cubic meter, Joules and degrees Kelvin) are extremely large, far greater than any conditions that we observe in our present Universe.

Paradoxically, the effective mass of a Planck string object is about 0.2 micrograms or the mass of a short strand of human hair. Such a mass is readily discernible and measurable by humans.

The sources for derivations and comparisons for these Planck entities are the fundamental constants appearing in the Standard Model for Particle Physics and the Heisenberg Uncertainty Principal in Quantum Mechanics. The Heisenberg principal asserts that there is a minimum magnitude for the size and energy state of certain physical parameters.

The following are examples of these relationships where "h" is Planck's constant with units of energy multiplied by time or electron volts (eV) times seconds. Planck's constant has the approximate value

$$h = 3.7 \times 10^{-34} \text{ eV-s}$$

It is important to realize how extremely small yet finite that Planck's constant is and its incompatibility for human perception. If "h" is expressed in ordinary energy units of Joule-seconds, we have that

$$h = 3.7 \times 10^{-34} \text{ eV-s} = 3.7 \times 10^{-60} \text{ Joule-s}$$

For instance, the Planck length at 10^{-34} cm is an entity that requires a sequential comparison as follows: Compare the size of an atom to the earth, then downsize this object again to the size of original atom and again compare the result to the size of the earth. The scale change for each of these sizing scenarios is 10^{-17}, and multiplying this size comparison results in the value of $10^{-17} \times 10^{-17} = 10^{-34}$ cm or approximately the Planck length. It is apparent that such a range in size is far beyond human experience or even comprehension.

M Theory critics

M-Theory is a bold, unconfirmed theory that has many critics who openly or indirectly reject M-Theory. These critics disparagingly label the single letter M in "M-Theory" as a pseudonym for derogatory epithets such as those in the following quotation that have appeared in a recent journal article lampooning M-Theory.

"Since the proponents of M-Theory cannot agree on the origin or meaning of the letter "M" in M-Theory, we, the old guard physics community composed of conventional, archaic physicists, have elected to assist our

iconoclastic, young colleagues with the following suggestions as to what 'M title' best describes for their new and improved brand of Physics. We request that the broad physics and scientific community of scholars and followers submit to the American Association for the Advancement of Science, or AAAS, for evaluation and consideration for awards for the most accurate and appealing entries. The following table of modifier candidates has been forwarded for edification and contemplation in addressing this dilemma facing our M-Theory colleagues."

<u>Modifier Candidates for the "M" in M-Theory</u>

Macabre, macho, macramé, mad, madness, mafia, magi, magic, maladaptation, maladroit, milk-of-magnesia (as a laxative), maladaptation, maladroit, malaprop,

malefic, malevolent, malice, malformed, maniacal, mass hysteria, maybe

Mean, meander, measly, meddlesome, mediocre, megalo, menace, mendacious,

menial, mercy, meretricious, mermaid, mess, metamorphosis

Mickey Mouse, mid-course, middling, midway, minor, miserable, mistaken

Mock, modicum, molehill, momus, mongrel, morass, morbid, moribund, morose, morph, mossy, motley

Mucid, muck, muddy, mulch, mundane, mush, mutagen

Myopia, mysterious, mystery, mystical, myth

Obviously, the above table forwarded by critics of M-Theory is meant to demean and possibly eradicate this cancer in the hallowed field of physics. The critic's particular bias for such unfavorable epithets is that M-Theory boldly claims that our entire universe and other possible "parallel or engulfing multiverse(s)" are imbued with the audacious property of ten spatial dimensions and one time dimension. M-Theorists assert that eleven dimensions are necessary and sufficient to unify the entire diverse world of physics from the quantum theory in the Standard Model of Particle Physics to the General Theory of Relativity.

M-Theory is further postulated by its proponents as the ultimate "Theory of Everything" or the so-called TOE for the complete mathematical basis that supports and explains the entire universal physical foundation of science. If M-Theory is the essential TOE, then it is the solitary appendage for the foot of physics that supports and governs all scientific knowledge. It is obvious that such an audacious claim and assertion by a small, renegade group of young, iconoclastic physicists is met not only with incredulity by layman, but with vociferous derision from the older, well-established community of mature and sage physicists. This larger and "wiser" population of physicists and scientists embrace and guard the bastions of ordinary spacetime physics that has dominated the conventional and sensible universe of four dimensions…three dimensions in space and one dimension for time for human history.

David M-Theory Leader

However, for M-Theory audiences and its limited, but dedicated proponents, David ben-Steinmann is the acknowledged leader and principal architect and spokesperson. David is the cause célèbre; and it has been through David's brilliant, but sometimes controversial, efforts that M-Theory is now studied, pursued, evaluated, criticized, and either reviled or embraced by physicists throughout the world. M-Theory is, indeed, very controversial; and the older and conventional physicists in the international physics community disparage such a claim for a "theory of everything." These critics assert that M-Theory is supported only by elegant, but superfluous, mathematics and obscure topological symmetries and group theory subtleties. Furthermore, without observable and confirming experimental data, M-Theory is an apparition or wishful philosophy with no credible evidence that is subject to experimental testing and confirmation.

Nevertheless, a small, dedicated cadre of scientists, mostly young nuclear physicists and a few prolific mathematicians, such as David, entertain and defend M-Theory as their single passion in physics. The passion that embraces and compels this small enclave of "young rebel physicists" is the resolute belief that M-Theory can and will successfully merge Quantum Mechanics and Einstein's General Relativity. A merger into a comprehensive and compelling theory that explains all physical phenomena observed and postulated in the known universe including bosons and hadrons, quarks and leptons to parallel universes and black holes that fill the known cosmos and possibly undisclosed universes in a multiverse occupied by dark matter and dark energy.

David's mathematics forte

As is to be expected, such an unorthodox theory as M-Theory has many critics and only a few stoic supporters. David Steinmann is one of these few, iconoclastic physicists who believe that they can demonstrate that M-theory is that final answer to the ultimate question, "What are the correct fundamental model and description for all physical objects and processes in the universe?"

David's particular expertise is advanced mathematics including analysis, complex variable theory, matrix and tensor analysis, and especially higher dimensional group theory. He believes that through the unique, abstract power of mathematics, he can demonstrate to the many critics of M-theory that higher dimensions and string entities are not only necessary and sufficient to merge quantum mechanics and Einstein-mechanic, but that eleven dimensions truly exist in the physical cosmos. David believes that mathematics does not suffer the myopia of human observations and limitations of anthropomorphic thinking. However, demonstration that M-Theory is valid is a formidable task and has been David's singular passion and pursuit in science.

David's Education

The main character in this science-fiction novel is David ben-Steinmann, the rising young mathematical physicist who leads the small army of scientists attempting to confirm their revolutionary theories of physics.

David ben-Steinmann was born in Jerusalem in 1980, the only son and second child of a Jewish mother, Sariah bin-Haidari, and a German father, Karl von Steinmann. David attended the Technion in Haifa, graduating with highest honors in theoretical physics in 1999 at the age of 19. He then traveled to the United States to enroll in a combined graduate program of physics and mathematics at the California Institute of Technology in Pasadena, California or more familiarly, Caltech. He graduated at age 23 with his doctorate or PhD, bestowed summa cum laude. A two-year postdoctoral tenure followed at Princeton's Advanced Institute of Studies and affiliate Universities set him on his present path in M-Theory at the age of 25. Because of his eminence in the field and value to the U.S. research community, his was granted early U.S. Citizenship also at the age of 25. He has served as a consultant to U.S. Department of Defense, Defense Advanced Research Projects Agency (DARPA), National Science Foundation, and the National Academy of Science to which he was elected a full member at age 28.

David's Position at Harvard

Now at age 30, David has garnered a lifetime-endowed chair at Harvard and joint Research Positions at the Institute for Advanced Study at Princeton University, MIT and other eminent U.S. and foreign universities because of his seminal work in theoretical cosmology. His consuming passion and the singular focus and goal in his research during his youthful life is the unequivocal verification of M-Theory and its adoption by the international physics community. David is driven by his obsession to unify the world of physics and finally merge the smallest structures of the atom and the largest structures of the galaxies and the universe.

David's ancestry

David Steinmann's ancestry is also unconventional. His grandfather Eric Von Steinmann was a Reich Colonel in the Third Reich in Nazi Germany during World War II. In a paradoxical insult to Nazi anti-Semitism, Colonel Von Steinmann found and rescued his future wife, a beautiful young Jewish woman imprisoned at Auswitz, the infamous Nazi Concentration Camp. The German Colonel, in an incredible saga of intrigue and danger, rescued his wife-to-be from certain extermination during the closing days of the Holocaust and the collapse of the Third Reich. He wooed and married this Jewish woman at the close of the World War II in Europe in April 1945. This unlikely wedded couple, at the insistence of his resolute, Zionist Jewish wife, eventually traveled to Israel to support the new nation of Israel that was also established under remarkable circumstances and recognized in 1948. In Israel, the Steinmann's established their home and reared their family in kibbutz, Sde Boker in the center of the Negev. Because Eric Von Steinmann's wife was Jewish and a close friend of David ben Gurion, the First Prime Minister of Israel, Eric and his wife were both able to claim Israeli citizenship immediately upon their arrival into Israel.

David's father, Karl von Steinmann was born in the kibbutz Sde Boker in 1950. He married Sariah Gurion in 1975. While serving in the military service in the Israeli Defense Force (IDF), he was recruited as an undercover officer in the Israeli Mossad in 2003, the clandestine organization for Israeli Intelligence. In 2005, David's father was killed in a secret, joint Israeli-American assault on an Al-Qaida cell in Lebanon. Because of the death of David's father in this controversial "friendly fire" incident, David's mother, Sariah, harbors enmity and rejection of all Gentiles and Americans in particular.

At the age of 18, Rebecca met and rebelliously married a Lebanese Arab named Mohammad Hussein, resulting in rejection of Rebecca's marriage by David's mother, Sariah. David's mother was infuriated at this unorthodox marriage, and punished Rebecca by refusing to accept or even acknowledge her Arab husband. Rebecca repeatedly attempted to find respite from her mother's animosity and solidarity for her fragile marriage. Unfortunately, the marriage did not endure its precarious heritage, as predicted and precipitated by her mother, Sariah. After Rebecca's divorce, she moved to America to escape her tyrannical mother. Rebecca is a now a nurse and lives in Belmont, Massachusetts with her brother David.

David was more compliant to his dominant Jewish mother's demands and pursued his undergraduate education at the Technion in Haifa. David's interests in science and mathematics and his unique mind and proclivity established him as a remarkable scholar who gained acceptance and favor of his widowed mother. However, after graduation from the Technion, David chose not to remain in Israel, and went to the United States to pursue his graduate education and research. He now lives in the United States with Rebecca in Boston, Massachusetts.

David's status in science

At age 30 as of 2010, David currently holds the Weisenhausern Endowed Chair in Theoretical Physics at Harvard. He is the youngest person at any academic institution to hold this esteemed lifetime position that he garnered because of his extraordinary achievements in physics and mathematics. The Board of Directors of the Weisenhausern Family Trust aspires to sponsor a Jewish Scholar in science who has the potential to become a Nobel Laureate. The Nobel Committee in Sweden had earlier nominated David for the prize in physics, but a split vote blocked the actual award to David. Several members of the Nobel Laureate Selection Committee demanded confirmation, preferably experimental, of David's advanced hypotheses to support his modifications to Super String Theory that resulted in "M-Theory." The Committee was reluctant to acknowledge M-Theory without a firmer basis for confirmation before granting the award.

David's Physical Description

David is tall at 6 foot 2 inches, lean, weighing 180 pounds and physically fit. He has an athletic build, dark hair, brown eyes and is an infrequent member of the dragon boat team at Harvard that competes in boat races on the Charles River during the summer. He could easily excel in any area of athletics, but his consuming passion is physics, so he reluctantly participates in vigorous activities only at the insistence of his colleagues and graduate students that he supervises.

David's epilepsy

Physically, David enjoys excellent health and great physical strength except for one serious and threatening medical issue...David is a confirmed, acute epileptic. Epilepsy is a genetic trait in the Steinmann ancestral line and David is a full recipient of this unfortunate, genomic legacy. He has experienced numerous "petit mal" or small seizures during his lifetime. Usually he can control these seizures with proper medication. At the age of eleven, David experienced a "grand mal" or great seizure that threatened his life. Fortunately, with timely and skilled medical intervention, he recovered. However, the impact of that grand mal seizure is branded in his memory and now constrains his personal driving activities.

Because of this latent peril, David does not drive his automobile more than a few kilometers alone from his home in Belmont, a suburb of Boston, Massachusetts. He confines his solo driving to moderate speeds and short distances around the Cambridge area. He is in constant fear of experiencing a grand mal seizure while driving. He would then be fully incapacitated and pose a risk to himself and others. Therefore, when David must travel by car a long distance, he enlists his sister, Rebecca, or another colleague to drive. Because of his dormant seizure syndrome, David depends upon Rebecca, a Registered Nurse, to provide him medical support. Furthermore, David's close colleagues and graduate students are aware of his epilepsy and are prepared to secure immediate medical attention for a serious seizure.

ABOUT THE AUTHOR

The author, Gary Sandquist, is Professor Emeritus at the University of Utah where he directed the Nuclear Engineering Program for over 3 decades. He has over 750 publications including a textbook titled, Introduction to System Science by Prentice-Hall. As a member of the nuclear science and engineering community, he is intrigued with the rapidly expanding field of cosmology and its revolutionary concepts that evoke divine design. This fictional novel is based upon his study of cosmology and his personal assessment of how this field of science may relate to the tenets of the Church of Jesus Christ of Latter Day Saints. However, the opinions of this author are solely his own and do not represent those doctrines of the Church or the prevailing theories of modern cosmology.

www.ingramcontent.com/pod-product-compliance
Lightning Source LLC
Chambersburg PA
CBHW071400170526
45165CB00001B/122